LEADERS AND LAGGARDS

NEXT-GENERATION ENVIRONMENTAL REGULATION

Neil Gunningham and Darren Sinclair

NEIL GUNNINGHAM is Professor in the School of Resources, Environment and Society, and in the Regulatory Institutions Network, Research School of Social Sciences, both at The Australian National University. He was previously Foundation Director of the Australian Centre for Environmental Law at the ANU, Visiting and Senior Fulbright Scholar at the Center for the Study of Law and Society, University of California, Berkeley, and Visiting Fellow at the Centre for the Analysis of Risk and Regulation at the London School of Economics. He is also a consultant to the Organisation for Economic Co-operation and Development (OECD), to the United Nations Environment Programme (UNEP) and to various environmental regulatory agencies in Australia.

Gunningham's research interests and publications are principally in the area of environmental regulation and safety, health and environmental policy. His previous books include *Smart Regulation: Designing Environmental Policy* (with Grabosky; Oxford, UK: Oxford University Press, 1998), *Regulating Workplace Safety* (with Johnstone; Oxford, UK: Oxford University Press, 1998) and *Environmental Outlook No. 3: Law and Policy* (ed. with Leadbeter and Boer; Sydney: Federation Press, 1999).

DARREN SINCLAIR is a Senior Research Associate at the Australian Centre for Environmental Law, The Australian National University. He has worked on a number of environmental regulation and policy research projects, and has been a consultant to several government agencies and industry organisations. Previously, he worked for the Australian Department of Industry, Science and Technology. Recent reports include *Environmental Partnerships: Combining Sustainability and Commercial Advantage in the Agricultural Sector* (with Gunningham; Canberra: RIRDC, 2002) and *Establishing Comparability and Equivalence amongst Forest Management Certification Schemes* (with Kanowski, Freeman and Bass; Canberra: AFFA, 2000). He has contributed to several books including Chapters 2, 3 and 6 (with Gunningham) in *Smart Regulation: Designing Environmental Policy* (Oxford, UK: Oxford University Press, 1998) and Chapter 7 (with Gunningham) in *Environmental Outlook No. 3: Law and Policy* (ed. Leadbeter, Gunningham and Boer; Sydney: Federation Press, 1999).

Leaders & Laggards

NEXT-GENERATION ENVIRONMENTAL REGULATION

Neil Gunningham and Darren Sinclair

Greenleaf
PUBLISHING
2 0 0 2

For Sue, Beth and Kate

For Allison, Chloe and Liam

Learning Resour
Centre

124 62535

© 2002 Greenleaf Publishing Limited

Abstracts originally published in *Regulatory Compliance Issues, Challenges and Best Practice in the Environmental Protection Field*. Copyright OECD 2001, final version of a report drafted by Professor Neil Gunningham.

Published by Greenleaf Publishing Limited
Aizlewood's Mill
Nursery Street
Sheffield S3 8GG
UK

Typeset by Greenleaf Publishing.
Printed and bound, using acid-free paper from managed forests, by Bookcraft, Midsomer Norton, UK.
Cover design by LaliAbril.com.

British Library Cataloguing in Publication Data:
  Gunningham, Neil
      Leaders and laggards : next-generation environmental
      regulation
      1. Environmental policy 2. Environmental protection
      I Title II Sinclair Darren
      333.7'2

      Paperback: ISBN 1874719489
      Hardback: ISBN 1874719497

CONTENTS

PREFACE

The origins of this book lie in our interactions with a number of very different organisations. First, it emerged out of our research partnership with two leading environmental agencies in Australia: the Victorian Environmental Protection Authority and the Western Australian Department of Environment Protection. Both agencies have attempted to find alternatives to their traditional regulatory strategies and indeed have already made considerable progress in doing so. However, they have been hampered by their lack of awareness of how similar agencies elsewhere are meeting similar challenges, and by a lack of data with which to evaluate their current initiatives. The development of a formal partnership, under the auspices of the Australian Research Council's 'Strategic Partnerships with Industry' initiative, facilitated our collaboration. Under it, we sought to identify regulatory best practice internationally in a number of specific contexts, to evaluate through fieldwork the effectiveness of regulatory reform in those jurisdictions and to provide policy prescriptions that would better enable them to fulfil their regulatory missions.

Second, Neil Gunningham's work for two branches of the Organisation for Economic Co-operation and Development (OECD) in Paris facilitated broader comparative research on regulatory reform across a range of OECD countries. Specifically, Gunningham was commissioned by the Public Management Service (PUMA) to report on innovative approaches to regulatory compliance internationally, and by the Environment Directorate to research the implications of voluntary approaches to environmental protection, focusing on the very different experiences of the mining and forestry sectors.

Through these latter projects it became very apparent that a lack of information about regulatory reform, of what works and what doesn't, and of how best to harness the resources of both government and non-government stakeholders, was not merely an antipodean concern. Indeed, it was precisely this concern, expressed by a number of OECD countries, which precipitated our work for PUMA. Progress is being impeded unnecessarily by a lack of shared knowledge and by limited fieldwork on the success or otherwise of existing initiatives.

In stating this, we do not wish to downplay some recent and extremely valuable attempts to engage with the problem. The activities of the European Union Network for Implementation and Enforcement of Environmental Law, of the US Multi-State Working Group (in relation to environmental management systems and regulation), of the North American Free Trade Agreement Commission for Environment Co-operation, and of the US National Academy of Public Administration, are all examples of successful information-sharing. But these are islands of wisdom in a sea of ignorance. For example, when it comes to dealing with small and medium-sized enterprises (SMEs) very little is known, and what is known is not effectively distilled and disseminated. Much the same can be said about many current initiatives to reconfigure the regulatory state.

The aspiration of this book is to complement existing research initiatives and to expand our knowledge of regulatory reform by showing how existing experience can best be put to practical use, by drawing lessons from experiments in innovative regulation internationally, by conducting our own case studies and extrapolating from them, and by advancing our understanding of which instruments and strategies are likely to be of most value and why. As such, and in contrast to some of our previous work (Gunningham and Grabosky 1998: Chs. 2, 6; Gunningham and Sinclair 1999a, 1999b), the intended contribution of the present project is more to the successful application of theory than to its development.

Our intended readership includes environmental policy-makers, regulatory and other government officials responsible for policy design and implementation, academics and postgraduate students in environmental management, environmental law and environmental policy, and a more general readership within environmental policy and management studies. The book will also be of interest to those in industry, such as environmental managers and corporate strategists, who are considering the use of more innovative environmental and regulatory strategies, and to environmental non-governmental organisations.

We are particularly indebted to the former chief executives of our partner agencies, Dr Brian Robinson and Dr Bryan Jenkins, for their personal support and for the commitment of senior agency staff more generally, to the project's goals, and for their invaluable feedback on our draft reports. We also wish to thank our many interview respondents who patiently answered our questions. Parts of Chapter 2 were first published as N. Gunningham, 'Regulating Small and Medium-Sized Enterprises', *The Journal of Environmental Law* Volume 14, Number 1 (2002). A modified version of Chapter 3 was published as N. Gunningham and D. Sinclair, 'Partnerships, Management Systems and the Search for Innovative Regulation in the Vehicle Body Shop Industry', *Business Strategy and the Environment* Volume 11 (2002), pp. 236-53. Parts of Chapter 4 first appeared in N. Gunningham, 'Environmental Regulation and Cleaner Production Partnerships with Small and Medium-sized Enterprises: A Case Study', *Environmental and Planning Law Journal* Volume 18, Number 4 (August 2001), pp. 369-80, reprinted with the expressed permission of the ©Lawbook Co., part of Thomson Legal & Regulatory Limited, www.thomson.com.au. Some abstracts will be published in *Regulatory Compliance Issues, Challenges and Best Practices in the Environment Protection Field*, copyright OECD, 2001 (final version of a report drafted by Professor Neil Gunningham) and are reproduced by permission of OECD.

1

INTRODUCTION

The environmental impact of industry, especially pollution, has been subject to regulation for at least three decades, under an approach that is somewhat unfairly called 'command-and-control' regulation. This approach typically specifies standards, and sometimes technologies, with which regulatees must comply (the 'command') or be penalised (the 'control'). And it commonly requires polluters to apply the best feasible techniques to minimise the environmental harm caused by their activities. Command and control has achieved some considerable successes, especially in terms of reducing air and water pollution (Cohen 1986; Cole and Grossman 1999). However, it has been widely criticised, particularly by economists, for inhibiting innovation, and for its high costs, inflexibility and diminishing returns (Sunstein 1990; Stavins and Whitehead 1997).

The problems of command-and-control can be overstated and its considerable achievements too easily forgotten (Cole and Grossman 1999). Nevertheless, its limitations have led policy-makers and regulators to recognise that it provides only a part of the policy solution, particularly in a rapidly changing, increasingly complex and interdependent world. As one US EPA (Environmental Protection Agency) official put it: 'as the world changes, patterns of law and governance must change with it' (Fiorino 1999: 467). And much indeed has changed over the past decade, with a plethora of innovative regulatory and quasi-regulatory approaches seeking either to replace or to complement command-and-control in a variety of contexts.

As policy-makers and social scientists observe this changing regulatory world, they have learned many lessons concerning: the limitations of command-and-control regulation; the potential value of a range of alternative regulatory strategies such as economic instruments, information regulation, regulatory flexibility and voluntary agreements (VAs); the development and application of more sophisticated enforcement strategies; and the value of cleaner production and environmental partnerships.

Yet there is still a great deal that is not known, or that is 'known' theoretically but not tested through fieldwork, or even that has been tested but which still does not inform regulatory practice 'on the ground'. In particular, our work with the Organisation for Economic Co-operation and Development (OECD) and our interactions with a variety of environmental regulatory agencies have sensitised us to the substantial extent to which the development of theory has outstripped its application,[1] and to the importance of synthesising and disseminating more effectively what is known about 'next-generation' environmental strategies.

These interactions have also made us acutely aware of the extent to which the translation of theory into practice must take account of shrinking regulatory resources. In some circumstances it is necessary to develop strategies capable of achieving results even in the absence of credible enforcement (as when dealing with small and medium-sized enterprises), and in almost all circumstances to extract the 'biggest bang' from a rapidly diminishing 'regulatory buck'. We believe this diminution in the capacities of environmental regulators is a mistake, and that considerable additional expenditure would be a wise investment in the public and environmental interest. However, we also recognise that, at least under prevailing political ideologies, and in the absence of large-scale environmental disasters, such expenditure is highly unlikely.

This book is about how to design regulation and alternatives to regulation in this economic and political context, in a manner that is both effective in protecting the environment and efficient in that it does so at least cost to regulators and regulated enterprises.[2] It takes as its starting point the proposition that command-and-control regulation has indeed made a substantial contribution in many areas of environmental policy, particularly in relation to laggards, and will continue to do so. However, it is also clear that 'the low-hanging fruit' has largely been picked, and that in an increasingly complex, diverse and interdependent society command-and-control is a blunt tool that is not well suited to meeting many of the challenges which lie ahead.

Our intention is not to rehearse the conventional arguments about the strengths and weaknesses of command-and-control,[3] but rather to examine the next generation of regulation and regulatory tools developed to curb the environmental excesses of business.

In doing so, we distinguish between the regulation of large enterprises and of small and medium-sized enterprises (SMEs). These two groups are treated differently, in different chapters, because for the most part the problems they face, their motivations, and, above all, the instruments most likely to be effective in improving their environmental performance, are quite distinctive. There is, for example, little point in using public shaming as a tactic on a small enterprise with no public reputation to protect, whereas the very same strategy can be extremely effective when used against large, reputation-sensitive enterprises.

We also distinguish between policies designed to bring laggards up to a minimum level of compliance and those designed to reward leaders for going 'beyond compliance'

---

1 For example, a recent OECD report on regulatory compliance concluded that: 'awareness of compliance problems is growing among Member countries, but that action to improve compliance is uncoordinated and unsystematic' (OECD 2000a: 8).

2 Effectiveness (contributing to improving the environment) and efficiency (improving the environment at minimum cost, within which we include administrative simplicity) are not the only possible goals of environmental policy. We also recognise the importance of equity (showing fairness in the burden-sharing among players) and of political acceptability (which includes factors such as liberty, transparency and accountability). However, effectiveness and efficiency are arguably the pre-eminent criteria, because in the majority of cases the effectiveness of regulatory policy in reaching an environmental target, and its efficiency in doing so at least cost, will be the primary concerns of policy-makers.

3 For a summary, see Gunningham and Grabosky 1998: Ch. 2 and references therein.

with the regulatory status quo. Traditional command-and-control regulation has focused firmly on the former group and on prescribing minimum levels of environmental performance, but rarely encouraged continuous improvement beyond that level. Yet, as we will argue, next-generation mechanisms can make a considerable contribution in improving the environmental performance of leaders as well as laggards and, in so doing, lift levels of environmental performance across the board. While strategies for encouraging beyond-compliance behaviour are much more advanced in relation to large enterprises (since the large majority of leaders will come from the ranks of such enterprises), such strategies can also benefit some SMEs.

The remainder of the book is organised as follows. Chapters 2–5 examine how best to overcome the considerable barriers to curbing the environmental excesses of SMEs. Chapter 2 provides a critical evaluation of instruments currently used for regulating SMEs, and builds on recent international experiments and our own work on small-business regulation to derive a set of policy recommendations for more effective and efficient regulation of this group of enterprises. In particular, it identifies policy mechanisms that can effectively harness the goodwill of SMEs and which are pertinent to their commercial circumstances, while at the same time retaining more coercive mechanisms to monitor and control laggards.

Chapters 3–5 provide greater depth of analysis through a series of case studies. These demonstrate how regulation of SMEs could be more effectively designed to: (1) facilitate industry self-management; (2) facilitate effective risk management; (3) foster environmental partnerships with larger firms (for example through the supply chain); (4) encourage the adoption of cleaner production; and (5) engage the financial sector in the environmental performance of SMEs. Our particular case studies relate to vehicle crash repairs, the vegetable-growing industry, and ozone protection in the refrigeration industry.

Chapters 6–8 examine the very different challenges involved in regulating large enterprises and how these might be overcome. Chapter 6 explores a variety of innovative forms of regulation. Not least, important lessons are derived from the 'Reinventing Environmental Regulation' initiative in the United States, from the European experience with voluntary agreements brokered by government against the background of less palatable alternatives, and from the various forms of informational regulation, which seek to harness both non-governmental organisations (NGOs) and financial markets in the environmental interest. More broadly, this chapter analyses a number of innovations, mostly in Europe and North America, that have a track record of success (and some that do not) in order to identify best-practice environmental regulation, as applied to particular circumstances.

Chapters 7 and 8 complement this broader synthesis with two case studies exploring the pitfalls of various next-generation regulatory mechanisms and demonstrate how our current regulatory regime can be substantially improved on the ground. Chapter 7 examines the experience of voluntary codes in the mining sector internationally, in order to derive broader lessons as to how to tailor such codes to best fit the circumstances of particular industry sectors, to identify their limitations and the extent to which they need to be supported by other instruments. Chapter 8 examines the Australian experience with Environmental Improvement Plans (EIPs) and accredited licensing, two very distinctive innovations which seek to achieve best practice through 'facilitative regulation'. Under this approach a combination of community participation, information disclosure and process-based regulation are used as mutually reinforcing means of driving industry performance up to and beyond compliance, and of encouraging leaders as well as pressuring laggards.

Finally, Chapter 9 locates the various next-generation regulatory reforms within a broader context. It argues that we are witnessing not a retreat of the regulatory state, but rather a reconfiguration in which the state's role has shifted from direct regulation to one of facilitation, empowerment and sometimes partnership, with a range of third parties and sometimes with business itself. It traces the main reasons why such a reconfiguration has come about and what it is likely to contribute to environmental protection, by examining recent reforms through a series of different conceptual lenses or perspectives.

The principal methodology used to conduct the fieldwork component of the project was semi-structured interviews with a representative sample of stakeholders, and in particular with business enterprises, government inspectors and a variety of other groups that also have a direct role in, and experience of, the regulatory process.[4] These included industry associations, environmental and community groups, and environmental professionals. This sample was supplemented by strategically targeted interviews with other key actors identified on the basis of 'snowball' sampling.

Our focus was on the design and application of environmental policy in the developed world and in countries of Anglo-Saxon origin in particular (we are less confident of the extent to which our conclusions might translate to other countries and cultures). Our fieldwork for this project was undertaken in such countries (primarily, but far from exclusively, in Australia), and the broader theoretical and empirical literature on which we draw in developing our arguments is also substantially taken from countries of Anglo-Saxon origin (principally Canada, the UK and the USA).

However, while many of the challenges facing environmental regulators in these countries are quite similar, their regulatory cultures are not. Like others before us, we would draw a distinction between the tradition of 'adversarial legalism' characterising regulation in the United States and the far more co-operative approaches taken in Australia, Canada and the UK (Kagan 2001; Kagan and Axelrad 2000). At the level of enforcement strategies, and in terms of establishing trust and partnerships between regulators and regulated enterprises, this distinction is crucial. Some of the conclusions from our case studies, all of which are based in the latter group of countries, will have more obvious and immediate relevance to that group than to the USA. Others among our arguments will, we hope, resonate for all of these countries.

As will be apparent, the book is principally normative in its approach. Its main concern is to identify weaknesses in current arrangements, and to argue the case for adopting a number of regulatory reforms and innovations drawing from experience internationally, or on recent and important developments in regulatory theory, and on the models and approaches we have constructed during the course of our own research. Accordingly, we hope that the analysis and recommendations of this book will benefit those who are interested in achieving better environmental regulation and policy, including environmental agencies and government officials responsible for policy design and implementation. We hope it will equally be of value to corporate strategists interested in improved environmental outcomes, NGOs and students of environmental law and policy.

---

4   A total of 138 interviews were conducted: 116 in Australia, and the remainder in the USA and Northern Europe.

REGULATING SMALL AND MEDIUM-SIZED ENTERPRISES

Small and medium-sized enterprises (SMEs) represent a very high proportion of all enterprises in industrialised economies. In the UK, for example, over 99% of all enterprises fall within this category (defined for present purposes as those with fewer than 200 people),[1] with about 90% of these enterprises having fewer than ten employees. SMEs have different environmental performance characteristics from their larger counterparts. In particular, they commonly have a higher level of environmental impact per unit, and lower compliance rates with health, safety and environmental regulation (Bickerdyke and Lattimore 1997). Although their individual environmental impact may be small, their aggregate impact may, in some respects, exceed that of large enterprises. For example, collectively, they are claimed to be the source of around 70% of total environmental pollution in the UK (Groundwork 1998; KPMG 1997). In recent years, the environmental impact of SMEs may have been compounded by a substantial increase in the number of such enterprises.

The effective regulation of SMEs is a substantial policy challenge for environmental agencies in all jurisdictions, not least because this group has a number of unique characteristics which inhibit the application of conventional regulatory measures.[2] These include:

- A lack of resources. This is exacerbated by higher compliance costs, a shortage of capital and economical marginality (Haines 1997).

- A lack of environmental awareness and expertise. Many are ignorant of their environmental impact, technological solutions to their environmental problems, or their regulatory obligations.

- A lack of exposure—including lower public profiles—which means that 'pressure groups gain none of the prestige, headlines and publicity by targeting

---

1 There is no single generally accepted definition of an SME. Different definitions are adopted by different countries and for different purposes. See e.g. Hillary 2000a.
2 For a general survey and documentation of the issues identified below, see the essays in Hillary 2000a.

SMEs that campaigns exposing the environmental misdemeanour of high-profile multinational bring' (Hobbs 2000: 150).

- A lack of receptivity to environmental issues. Many SMEs have not integrated environmental issues into their business decisions, making it difficult to persuade them of economic benefits (Merritt 1998).

- The sheer numbers of such enterprises. This leads to infrequent inspections, and many businesses slip through the regulatory net and are untouched by environmental policy initiatives.

What motivates SMEs and their owners is also likely to be very different from what motivates large corporations. Most striking in this regard is not only their low level of environmental awareness but the substantial disparity between their environmental aspirations and their environmental performance. A 1999 study, for example, suggested that the typical owner-manager suffered a low standard of eco-literacy and poor environmental awareness (Tilley 1999: 241) and that while SMEs commonly expressed pro-environment attitudes they often experienced difficulties translating these ideals, aspirations and values into action (Tilley 1999; Hutchingson 1993).

The problems of ignorance and lack of awareness of regulation are even more severe in the case of very small enterprises or 'micro-businesses' (under five employees). In the closely related area of occupational health and safety (OHS) regulation, it was found that the vast majority of micro-businesses did not know that any legislation existed and were unaware of reliable advice sources. Most did not belong to any industry associations and were focused primarily on economic survival. Where it existed at all, concern about safety, health and the environment was limited to quite specific overt threats (Eakin 1992). Consistent with the above evidence, a Groundwork report (1998: 11) concluded that:

> a typical SME is ill informed and unwilling to take action unless threatened by strong external forces such as prosecution or customer demands. *Worse still, many foresee no threats or advantages to their companies from the environment* [emphasis added].

The policy challenge is made even more complex by the fact that SMEs are a very diverse group of enterprises, both within *and* across different sectoral groupings. And these variations may not only impact on the degree and type of environmental problems confronting a particular business and/or SME sector but also on the way in which an individual business and/or sector might be regulated or otherwise encouraged to improve its environmental performance. For example, despite the significant disadvantages of a lack of resources, information, economic security and external pressure, *some* types of SME are capable on occasion of responding with great flexibility and innovation.[3]

3   Two sectoral examples of this phenomenon are: SMEs working in innovative and fast-moving high-technology areas where high environmental performance may provide a competitive advantage; and those that perceive the risk of health, safety or environmental failures to be high. This latter group (for example, those working with high-risk chemicals) may be intrinsically motivated to improve without external prompting. However, these examples are likely to be the exception to the rule, with the very large majority of SMEs unlikely to respond voluntarily because they perceive no threats or advantages in doing so. For these firms, improvements are likely to come about only through the imposition of outside pressure (Wright 1998).

Recognising these problems, how can policy-makers overcome the considerable barriers to improving the environmental performance of SMEs, and design a strategy for their efficient and effective regulation? How can they bring the large majority of SMEs into compliance with environmental regulation? How can they persuade them to integrate environmental considerations into their core business activities? And, considering the diversity of SMEs, how can they develop policies that bring laggards up to the minimum legal standard while encouraging and rewarding leaders for going far beyond it? Unfortunately, SMEs are a particularly challenging subject for research. For example, one survey of 875 SMEs received only 15 responses, illustrating the understandable unwillingness of already under-resourced enterprises to co-operate in such research (Gunner 1994). In consequence, the existing literature on these issues is extremely sparse, and the fieldwork picture is extremely patchy.

Notwithstanding this disappointing track record, we believe it is possible to address the question of effective SME environmental performance, and at least provide provisional solutions to many of the regulatory challenges confronting SMEs. In doing so, it is important to recognise that, just as the motivations, attitudes and circumstances of SMEs may differ substantially from those of their larger counterparts, so also will the 'pressure points' that impact on enterprises' behaviour. Accordingly, an effective strategy must be tailored to their special characteristics.

This section of the book has two principal goals. First, in this chapter we seek to specify a series of instruments that are capable of substantially influencing the attitude and behaviour of SMEs and of engaging both leaders and laggards. We do so by drawing from the international literature on best-practice regulation (Gunningham et al. 1996: 60), and on our own fieldwork in Victoria and Western Australia (described below). From this, we identify the advantages and limitations of particular instruments and how each may be best designed to achieve improved environmental performance on the part of SMEs. We do not purport to address the entire range of regulatory instruments, but only those with particular capacity to influence SMEs' environmental performance and only those that might credibly be used within existing resource constraints (i.e. those that give the 'biggest bang for the regulatory buck').

Second, recognising that one size does not fit all, and that different industry sectors have quite different characteristics and require different instruments and policy mixes, we seek to develop an industry-specific strategy for improving the environmental performance of SMEs. We do so in outline at the end of this chapter, and in detail in a series of case studies in the following chapters. While it is not practicable to examine more than a representative sample of SMEs in this way, we are able to identify a range of SME sectoral attributes that might influence how the policy instruments described in this chapter are best applied in particular instances.

In engaging in these tasks it must also be recognised that all instruments have both strengths and weaknesses and that none is likely to be wholly successful in achieving its environmental goals. As we have argued in our previous work (Gunningham and Sinclair 1999a; Gunningham and Grabosky 1998: Ch. 6), it is often the combinations of instruments, and the interactions of various institutional actors, including governments, commercial and non-commercial third parties, that determine the success of environmental regulation. This suggests a hierarchy of instruments. For example, some instruments are best used initially, and only to the extent that they fail will it be appropriate to invoke other, more interventionist, approaches. In contrast, in other circumstances, combina-

tions of instruments should be used concurrently. A sequenced approach, gradually escalating from more co-operative to more interventionist instruments, may not only make
the best use of scarce regulatory resources but also better motivate target groups. Finally,
different industry sectors may have quite different characteristics and therefore may
require different instrument mixes. This strengthens the case for developing industry-
specific strategies.

In short, this chapter seeks to synthesise and build on the very limited and fragmented
international literature on environmental policy instruments targeted at SMEs. It integrates this with our own research and it seeks to construct an overall framework for how
best to deal with what has so far represented an intractable environmental challenge.
While for the most part it focuses on mechanisms that might bring environmental laggards up to acceptable levels of environmental performance, some instruments (especially environmental management systems and cleaner production initiatives) are more
ambitious, and seek to encourage continuous improvement and to reward environmental
leadership.

Education and training

Ignorance and a lack of capability are common explanations of poor environmental
performance in SMEs.[4] Beyond a limited understanding of their regulatory obligations,
the cleaner production literature suggests that SMEs are simply unaware of many
financially attractive opportunities for environmental improvement. This is compounded
by a shortage of technical and management expertise. One conclusion is that the large
majority of SMEs simply 'do not possess the knowledge, skills or solutions necessary to
allow them to fully integrate the environment into their business practices' (Tilley 1999).

These problems are exacerbated by a number of attitudinal obstacles on the part of
SMEs to improving their environmental performance. These include: underestimating
the impact of their activities on the environment; a narrow view of the relationship
between business performance and the environment; the entrenched idea that protecting
the environment is associated with technical complexity, burdens and costs; and a high
resistance to organisational change (Gunningham and Sinclair 1997). Although these
problems are not unique to SMEs, they are most prevalent within this group. Consequently, they have to be overcome before SMEs will be willing to improve environmental
performance.

Unlike larger enterprises, most SMEs lack the internal resources and motivation to
overcome the environmental challenges that confront them.[5] For this reason, there is a
strong argument for providing information and education to SME executives and owners,
tailored to their specific needs, which seek to modify their attitudes and behaviour.
Although this proposition is not in serious dispute, achieving such change is problematic
(UK Round Table on Sustainable Development 1999; KPMG 1997). While there is some
limited evidence of SMEs switching from end-of-pipe to cleaner technology, many well-

---

4   Parts of the following analysis first appeared in Gunningham 1999.
5   Indeed, it is arguable that regulators lack sufficient technical expertise to provide specific
    advice even to larger business.

intentioned efforts by governments and industry alike to promote the competitive advantages of cleaner production, for example, have had little impact on SME behaviour (Merritt 1998). The evidence suggests that there are considerable difficulties in persuading SMEs to act on environmental information, even when it is demonstrably in their own financial interest and/or backed by generous financial subsidies (Merritt 1998).

For example, in the UK, the Department of Environment's Small Company Environmental and Energy Management Assistance Scheme (SCEEMAS) provides a 50% subsidy for the costs of consultancy fees in the implementation of the European Union's Eco-management and Audit Scheme (EMAS). Despite a comprehensive national advertising campaign, and supporting material such as case studies, guides, videos, newsletters and leaflets sent to thousands of SMEs, a subsequent review revealed that only 136 individual SMEs had participated in SCEEMAS (the scheme was subsequently abandoned) (ECOTEC 2000).

Yet not all education and informational initiatives have been unsuccessful, and much depends on how the information is presented and packaged, and on who presents it.[6] Drawing on what limited empirical literature is available, it would appear that a number of issues are crucial to successful policy implementation. These are:

- Capitalising on win–win solutions. The starting point for effective communication, information dissemination and education should be to focus on those circumstances where good environmental practice can also be good business practice[7] and to emphasise that what is good in environmental terms may also be good for the economic bottom line (APEC 1999).

- Developing industry–government partnerships. The aim of such partnerships is to actively engage an industry in the development of a cleaner production strategy that is tailored to its particular circumstances. This generates ownership, thus increasing awareness and the level of commitment to its implementation, and emphasises improved environmental management practices.

- The right people disseminating the information—which must not only be transmitted, but also *received*. This is most likely to be achieved where there is face-to-face distribution from trusted sources (customers, suppliers and competitors, industry peers, networks and associations) which emphasises practical solutions. Information should also be sector-specific, and delivered in a co-ordinated fashion. The various forms of information delivery must be effectively co-ordinated, preferably by government, to minimise duplication (Fanshawe 2000).

- Developing codes of practice. SMEs often require much more specific guidance on what is required of them than their larger counterparts. Codes of practice are an effective way to provide practical guidance on how to achieve com-

---

6 A recent survey points to EMS support schemes in Germany and the Cleaner Production Programme in the Netherlands as examples of successful initiatives that included substantial information-based components. See ECOTEC 2000.

7 For example, it has been noted that adopting EMAS brought savings within only 14 months and that packaging reduction and re-use also saved considerable sums for many companies. See ECOTEC 2000.

pliance, and may be a valuable vehicle for promoting appropriate cleaner production benchmarks within an industrial sector.

- Exploiting third-party leverage. Most SMEs have frequent interaction with professionals (banks, lawyers, insurance companies) and larger enterprises along
the supply chain, and rely on them as credible sources of information. This
provides opportunities for using such professionals both to disseminate information and to exert pressure on SMEs to pursue opportunities for using
environmental improvements to achieve greater business success. On the basis
of enlightened self-interest (backed up by government persuasion), accountants might verify rudimentary environmental audits, banks might require an
environmental checklist for loan approval, insurers might seek a statement of
hazards identification and control, and larger enterprises may impose environmental management system (EMS) requirements (Hopkins 1995).

- Using more active 'hands-on support'. The ECOTEC report on SMEs emphasised
the importance of providing continuing on-site help over months or years
(rather than just 'self-help' information, limited training or brief environmental reviews), and noted that 'environmental graduates, through industrial
placement schemes, can be used to provide extended support at low cost, bridg-
ing the gap between existing staff, who have the company specific knowledge
but no time, and the student who has the time and the generic environmental
knowledge' (ECOTEC 2000).

- Integration with other strategies. Information and education cannot be relied
on to influence SME behaviour in isolation (Merritt 1998). They must be seen as
one component of a broader, integrated preventative strategy. What is needed
is a hierarchy of controls, beginning with the facilitation of voluntary action
through the dissemination of information and advice and support for cleaner
production initiatives, escalating through the use of positive and negative
incentives to encourage those who are otherwise constrained from taking
preventative action through cost or other constraints, and culminating in the
enforcement of command-and-control legislation for recalcitrants who are
unpersuaded by less interventionist strategies. However, information and education are almost always a necessary base from which other, more interventionist, instruments can be launched.

Facilitating self-inspection and self-audit

Given the limited resources of most regulatory agencies, and the limited reach of many
conventional regulatory strategies, there is a need to shift away from direct regulation
towards a variety of alternative strategies, involving voluntary compliance, self-assess-
ment and the use of third parties as surrogate regulators. While some such strategies are
best suited to large business (and are not further examined in the present context), others
are particularly appropriate to the environmental performance of SMEs.

Two instruments with considerable potential in this context are self-inspection and self-audit. These are less comprehensive and ambitious in scope than EMSs (which are discussed in a separate section below). For example, they seek regulatory compliance as opposed to continuous improvement, and generally apply to a more limited range of issues than EMSs. There are an increasing number of examples of this approach both in the case of environmental protection and related areas. The Queensland Division of Workplace Health and Safety, for example, provides audit documents that enable enterprises to self-audit and to illustrate how safety risks can be identified and controlled, thereby encouraging all organisations to take greater internal responsibility for risk management. By publicising in advance key audit criteria, workplace compliance may be facilitated even without inspections (Queensland Department of Employment 1994). In some sectors, workplaces (particularly SMEs) are required to complete Workplace Health and Safety Plans to improve their understanding of, and capacity to meet, regulatory requirements. Significantly, this encompasses principal contractors, contractors and subcontractors.

In Western Australia, a somewhat similar approach has been taken in controlling the disposal of waste solvents. Commercial users of solvents are required to register with the regulatory agency and pay a modest fee. They must also submit a short annual report to the regulator identifying how much solvent they use, what quantity of waste solvent is generated and to describe their waste management plan. It is possible to identify reports that fall outside industry norms, the providers of which may then be subject to an external audit. While this system of self-reporting both saves scarce regulatory resources and provides information necessary to identify potential recalcitrants, it will not identify calculated and sophisticated non-compliers that provide plausible figures to the agency.

One of the most advanced approaches to promoting voluntary self-inspection and self-audit among SMEs has been introduced in Minnesota through the Environmental Improvement Act of 1995. Under this statute, SMEs are encouraged to self-inspect, and to report the results to the state regulator, by being offered (limited) statutory protection from enforcement action. An additional incentive for participation is a 'Green Star' award for firms that do a complete environmental audit or inspection.

This approach has been further developed in the printing industry, where an agreement has been reached between the state and the local industry association to significantly increase the use of environmental audits by printing firms, many of which are relatively small operations. Under the scheme, the Printing Industry of Minnesota Inc. (PIM) established a separate corporation, PIM Environmental Services Corporation, to provide auditing services to PIM members. The auditor provides a complete evaluation of all a company's environmental, safety and health systems and submits the report to the company, and the company submits to PIM its plan for correcting any compliance problems. PIM then takes on an enforcement role in holding the company to its plan, throwing it out of the scheme if it fails to comply.

Although it was necessary to provide some incentive to firms to engage in such audits, government regulators were reluctant to offer a total amnesty from prosecution for breach of regulations, for fear of damaging the integrity of their enforcement programme. To overcome this problem, a compromise agreement was reached: an enterprise that audits itself, and which discovers environmental violations and corrects them promptly, will have this fact taken into account when regulators decide whether to initiate any enforcement action, whether an enforcement action should be civil or criminal in nature, and what penalties to impose. Thus 'a company which conducts an auditing program in good faith and makes appropriate efforts to achieve environmental compliance is likely to mitigate the consequences of any violations it discovers' (Humphrey 1994: 1).

However, notwithstanding the promise of the scheme in providing a credible, flexible and cost-effective alternative to command-and-control, its take-up by small printing firms has been less than anticipated because 'most printers see the costs of hiring the auditor and correcting problems as high compared with the low risk of being inspected by the state' (NAPA 1997: 134). As such, the presence of a potent 'enforcement stick' for non-participants, in conjunction with self-audit incentives, may be more effective.

Just how an appropriate combination of carrots and sticks might be developed is illustrated by the contrasting experiences under the Minnesota Act of the regulation of auto-body and repair shops, and of underground storage tanks. In the case of the former, regulators wrote to 4,400 auto-body and repair shops notifying them of the audit programme and urging them to use an enclosed self-inspection checklist, but did not threaten to inspect the facilities. The response rate was a modest 4.6%. In the case of underground storage tanks, a contrasting approach was tried. Letters were sent to the owners of these tanks notifying them of the audit programme but also threatening inspection for those who did not participate in the programme. The letter began:

> As part of the Minnesota Pollution Control Agency (MPCA) Underground Storage Tank (UST) program, I will be inspecting the USTs at one or more of your UST facilities during the next six months . . . The MPCA is giving you advance notice of our intent to inspect one or more of your UST facilities to give you the opportunity, if necessary, to bring your USTs into compliance.
>
> As an *alternative* to facility inspections, the [agency] has implemented a new self-audit program which enables owners of facilities . . . to self-evaluate their facilities to determine their state of compliance. If it is determined through the self-audit process that your facility is not in compliance with MPCA rules, you have 90 days . . . to bring your facility into compliance. Violations disclosed and corrected as a result of the self-audit will not be subject to fines or other penalties . . . [emphasis original].

A far higher response rate was achieved under this approach, whereby SMEs are given a clear understanding that the choice is 'not between compliance and non-compliance but between a low-cost, low-stress, collaborative route to compliance on the one hand and fines, liability, and public notoriety on the other' (NAPA 1997: 1,270). The relatively poor response of the auto-body and repair shops can be attributed to the lack of a credible threat of inspection and enforcement.

The underground storage tank approach may be an effective model to follow in terms of reaching SMEs that have gone relatively unregulated in the past. It also provides a pertinent example of the importance of establishing the correct balance between enforcement and assistance:

> Without maintaining a credible threat of enforcement [the regulator] lacks the leverage to get small businesses to invest in self-monitoring, let alone compliance. Without establishing an attitude of assistance and forgiveness [the regulator] will be unable to win the trust of small business owners, and those owners will be unwilling to accept the technical assistance they need to identify and correct their problems. A strong education and outreach program has

successfully linked those levers . . . to make the system even more credible, [the regulator] must begin conducting spot checks of firms submitting audit and self inspection reports to determine if companies are doing a sufficiently good job on their self-inspections and corrective action and to demonstrate the risk of false reporting. Of course, the agency also needs to make regular inspections of facilities that have not participated in the self-inspection program and to ensure that its entire regulatory program is credible (NAPA 1997: 134).

Finally, the importance of additional 'carrots' to reinforce the promise of audit protection should not be discounted. However, the current incentive, the Green Star awarded to participating enterprises, is failing to achieve its potential. In order for it to have meaning to a firm's customers and become an incentive for the firm, state agencies must publicise the award directly to consumers of services such as gas stations and print shops. Arguably, the Green Star could also be made a factor in the procurement of commodities, goods and services, although it is possible that this could lead to some confusion with other eco-labelling schemes.

A systematic approach to SME environmental performance

EMSs hold out the promise of achieving continuous improvement and cultural change in firms' approach to the environment. In particular, they have the capacity to integrate production with environmental considerations, to improve competitiveness and to increase profits. The conventional wisdom has been that EMSs are more suited to large businesses, with their complex operations and sophisticated management processes. Increasingly, however, it is recognised that there are extensive benefits to be gained by SMEs that adopt formal EMSs (Hillary 2000b: 144). Yet it is likely that SMEs will respond differently from larger businesses to the prospect of adopting EMSs, and that a different type of EMS will be suited to their circumstances. The challenge for regulators is to tailor EMSs, both in terms of their content and delivery, to the particularities of SMEs.

There are several obstacles to SMEs adopting an EMS (including the international standard EMS, ISO 14001), not least of which are a lack of resources (time, people and money), a lack of knowledge and technical capacity to adopt such systems (there are invariably no environmental specialists in SMEs), the fact that most of the costs are upfront and the benefits long-term, and a lack of commitment and market pressure to do so (related to lack of perceived benefits).[8] As if this were not enough, Hillary documents how 'negative company culture towards the environment and the disassociation between positive environmental attitudes of personnel and taking action cause the uptake of environmental performance improvements and EMS adoption to stumble at the first

---

8   The present environmental management standards are 'market-based' instruments which rely on the assumption that the market will reward companies for participating. Thus 'if there are no immediate market rewards—which appears to be the case for many SMEs—the instruments are flawed as mechanisms to improve environmental performance for 99.8% of UK businesses' (Gerstenfeld and Roberts 2000: 115).

hurdle' (Hillary 2000b: 144). The net result is that, although in theory SMEs might appear likely to embrace such systems voluntarily as a matter of enlightened self-interest, in practice this is far from the case.

An additional hurdle, and arguably the most serious barrier of all, is that the costs of accreditation are seen as disproportionately high for SMEs. According to some estimates, these are around 1.7% of sales for an SME with sales of US$500,000 per annum (Johannson 1997), with costs relating to consulting, documentation and training in addition to the costs of registration. Finally, only companies with more than about 50 employees tend to have formal management systems of any kind, with SMEs with under 50 employees tending to exhibit an informal style of management, which makes the use of formal management systems much more challenging (Johannson 1997). This was put graphically by one commentator as follows:

> The 'accounting system' is a plastic bag; when you get sophisticated about it you use a shoebox, really sophisticated is a banker's box. 'Communication systems' means sticky notes are left on the wall, phone, computer screen, and/or on the door for the person you wish to communicate with (Johannson 1997).

The lack of sophistication of SMEs has led some commentators to question whether formal EMS accreditation is appropriate (although work is progressing on this [ENDS 1999a]). The answer is likely to be a cautious 'yes'. Early evidence suggests that it is indeed possible to develop simple, certifiable management systems, capable of being adopted and used successfully by SMEs. For example, German firms have demonstrated that the comparable EMAS is not beyond the reach of SMEs, and in the UK a scheme has been introduced to encourage small firms to introduce EMSs with the support of the European Commission and national firms wanting to green their supply chains.[9]

Similarly, Per Langaa Jensen's research on Danish firms shows that many SMEs are capable of conducting systematic activities for the work environment, even though most choose not to do so. Moreover, the evidence suggests that: 'despite the problems, SMEs do find advantages in adopting EMSs. These include "spin-off" management benefits such as better organisation and business efficiency, as well as financial savings and improved environmental performance' (ENDS 1999a: 5). However, 'Companies often need outside support and a supporting surrounding structure' (Work Life 1999).

Arguably, the key, at least for SMEs with fewer than 50 employees, is to focus on simple, accessible improvements in management practices, rather than the introduction of formalised, administratively complex EMSs. For a five-person firm, for example, an EMS may be recorded in a few pages, or by using software tailored to individual SMEs and avoiding substantial documentation. Pursuing an even more informal approach, the New South Wales WorkCover Authority is encouraging SMEs to develop their own risk management procedures.[10] Here, references to formal systems are avoided, and instead a management approach develops uniquely in each particular workplace, accommodating the needs, expectations and responsibilities of individual workers.

---

9 The Business Environment Association, established to help SMEs improve their environmental performance, has launched an environmental benchmarking system aimed specifically at SMEs, and offers a stepped approach to meeting EMAS or ISO 14001 requirements. See further ENDS 1999a: 5.

10 www.workcover.nsw.gov.au

The pressure to maintain liquidity, the emphasis on short-term profit, and the pressures resulting from economic marginality all militate against the voluntary adoption of EMSs. How then, might SMEs be encouraged to overcome these barriers? One potentially powerful driver is supply chain pressure. We deal in detail with this issue below, noting that, while survey evidence suggests that very few SMEs currently perceive any pressure from their customers or suppliers to improve environmental performance, in a few industries this is changing and there are good prospects for making much greater use of this mechanism in the long term.

Another potential driver is government, which can play three key roles in encouraging the use of EMS by SMEs. First, it can provide information and education. For example, in Canada, recognising that the biggest obstacle to SMEs adopting ISO 14001 is that they have not even heard of it, the most important contributions have been:

* A recorded question-and-answer phone system accessible to anyone interested in generic information regarding EMSs

* A web-based learning tool for SMEs on EMS which breaks down the key elements of an EMS aligned to ISO 14001 into easy pieces in a matrix (Johannson 1997)

A role for government is the provision of external subsidies to help cover the cost of EMS adoption by SMEs.[11] In theory, there are objections to using subsidies for such purposes. In economic terms, it is important to achieve 'efficient' resource allocation by ensuring that firms 'internalise externalities', thereby ensuring that the costs of environmental degradation are borne by the person or enterprise that causes them rather than passed on to the taxpayer. However, this is one area where pragmatic considerations might justifiably prevail. We know that the start-up costs of developing an EMS approach are considerable, and that in the short term such systems are beyond both the financial and technical means of many SMEs. Accordingly, it may well be in the public interest to provide seed money to 'kick-start' a systems-based approach in smaller firms (or to 'buy out' the time of managers so that they can attend EMS training workshops).

However, whether or to what extent SMEs would respond to subsidies and other incentives remains an open question. The take-up of such initiatives in both the UK and Minnesota has been low, and the subsidy involved may need to be quite substantial before it has much impact. This would put a substantial, and probably unacceptable, financial burden on the state. Certainly the use of such subsidies on a large scale is likely to prove prohibitive (Johannson 1997).

A more modest alternative would be to pursue a proposal by Zeimet et al. (1997) for a management system approach tailored to individual company circumstances. Recognising that SMEs do not have the appropriate expertise available to facilitate OHS (or environment) measures, they offer a simplified structure that will assist them to do so and detail in particular the steps that need to be included in an employer action plan. In their view such a programme would include: employer and employee endorsement of the plan; a good vision statement; designating duties and responsibilities for those involved; a compliance policy; the identification, evaluation and control of workplace hazards; facility maintenance; ongoing accident investigation and record analysis; employee information

---

11   This subsection first appeared in Gunningham and Sinclair 1997.

and training; emergency preparedness; and an annual programme audit.[12] Again, it would be necessary to combine such an approach with a system of incentives and an underpinning of deterrence to make it credible. Box 1 describes one approach to designing a streamlined EMS for SMEs.

Solutions should be:

- **Inexpensive and sensitive to the limitations of SMEs.** Although SMEs are diverse in nature, one commonality is their lack of financial security and ready access to cash. It is essential therefore that environmental policies, including the use of management systems, take account of this characteristic by being affordable. Financial subsidies may also be required.

- **Co-operative.** Policies in respect of SMEs should not only seek to develop a spirit of co-operation between regulators and SMEs, but should also be based on co-operation between participating SMEs (and other relevant parties).

- **Locally based.** The issue of sustainability often benefits from the pursuit of locally based policies and strategies. In the case of SMEs, they are usually locally owned, use local employees and trade locally. It follows therefore that environmental improvement programmes should follow and exploit this pattern, and that such an approach is likely to generate the greatest leverage among other local SMEs.

- **User-friendly.** In essence, this means translating esoteric and complex environmental improvement strategies into easily understood terms, highly practical guidelines and readily achievable outcomes. Excessive use of bureaucratic and/or legalistic jargon is a particular hurdle in this regard.

- **Flexible.** For optimal results, SMEs should not only be able to tailor responses to their particular needs and circumstances but also to dictate the pace at which they implement changes. This requires flexible programmes and policies. A rush by authorities to impose both the timing and/or nature of environmental improvements will often lead to the adoption of ad hoc end-of-pipe solutions that lack management commitment and the potential for continuous environmental improvement.

Box 1 **EMS solutions for SMEs**

Source: Gerstenfeld and Roberts 2000

---

12 While such an 'action plan' may not be substantially different from a formal (albeit simplified) EMS, this approach has the virtue of not intimidating SMEs about the formality or scale of what is required of them. This is, for example, the approach of the New South Wales WorkCover Authority, which encourages employers to take certain steps without imposing any formal 'systems-based' approach on them.

Buyer–supplier relations

In many sectors there are massive disparities of commercial power along the supply chain that can be harnessed in the interests of environmental protection. Larger enterprises, in particular, may be able to impose product and process preferences on other enterprises, using their market power to influence the behaviour of upstream suppliers and down-stream buyers.[13] A 1998 UK research study documents the increasingly important role large enterprises are playing in recruiting and controlling their smaller trading partners (Rimington 1998). Rimington explains that big enterprises are simultaneously *reducing* the number of their suppliers and *increasing* their demands on them—insisting on higher standards of management and higher standards of safety, health and environmental performance from those they do business with. He notes that this is not usually for ethical reasons but rather for business reasons, connected with a distrust of commercial partners whose management standards are suspect in any way, and because this will increase the risk of loss and liability incidental to such partnerships.

Supply chain pressure thus offers a valuable means of influencing the environmental behaviour of SMEs. And, given the difficulties government faces in regulating SMEs directly, it may prove to be an important and effective complementary strategy. As Rimington (1998) asserts: 'co-operation and commitment of large companies in working with small firms—whether contractors, suppliers or neighbouring businesses—is vital in bringing about real improvement'. Importantly, a number of other studies also report a high level of compliance with safety, health and environment requirements where these are addressed under customer-dictated schemes. As Wright (1998: 19) points out:

> in these situations organisations appear to accept requirements as an unavoid-able and necessary condition of doing business, and will proactively comply. Where their requirements are dictated by a client (and applied with rigour) they are accorded equal status and importance as other client expectations.

An increasing number of approaches to supply chain relations have evolved in recent years, probably the best known and most advanced being the chemical industry's Responsible Care product stewardship code of practice. Under this code, larger enterprises have taken steps to influence the behaviour of SMEs through the practice of product steward-ship or 'cradle-to-grave' policies. These entail taking corporate responsibility for the life-cycle of a product, from the extraction and consumption of raw materials, through its manufacture, use and final disposal.

Some of the most advanced supply chain initiatives relate specifically to EMSs. For example, in Australia, the Ford Motor Company requires its several hundred auto produc-tion suppliers to be certified to ISO 14001, and it is helping them meet the standard through workshops and a group-oriented programme. However, it is also making it clear that, unless suppliers are prepared to be certified under ISO 14001, the company will not be able to commit itself to future orders from them. Toyota also seems likely to assist suppliers in developing and implementing an EMS, while Dulux, as a paint supplier, has similarly had an important role in introducing EMSs to the smash repair industry (Clay 1998).

---

13 Given that supply chain pressure depends on disparities in commercial power, it is unlikely that SMEs would be in a position to exert this over other SMEs, or indeed large businesses.

A recent variation on contractor pressure has been developed by the AGL Gas Network, which now requires major contractors to implement and maintain a health, safety and environmental management system. Importantly, the performance of these contractors is continually measured against set criteria. Part of the contractor's payment is dependent on overall performance, measured by ongoing field audits, thereby providing an ongoing financial incentive. By the project's second year, there was a marked improvement in safety, health and environment performance.

The motivation for larger enterprises to employ these initiatives is complex. It includes risk minimisation (a supplier closed down for poor environmental performance could both disrupt the supply chain and cause serious reputational damage to its trading partner) and cost savings from more efficient production practices, including reduction in waste disposal. In the case of the AGL Gas Network, for example, it gets a better-performing distribution network which more than compensates for any increased costs in rewarding high-performing contractors, while improved performance by contractors makes them both more efficient and more profitable (Stromvag 1998).

While such initiatives may be applauded, it should be noted that at present only a very small minority of enterprises are imposing such requirements on their supply chain partners, and that the vast majority of SMEs report no pressure from their customers to improve their environmental performance. As Hillary (2000b: 144) notes:

> Customers are a key driver for the adoption of EMSs and have influence far beyond any of the other stakeholders . . . Paradoxically, customers also show lack of interest in, or are satisfied with, SMEs' current environmental performance. Micro enterprises, in particular, found their customers to be uninterested in their environmental performance.

More broadly, there are a variety of roles that government can play in encouraging, facilitating and rewarding large enterprises to be more proactive in exerting pressure on the SMEs who are their customers. It might, for example:

- Exert its own supply chain pressure through its procurement policies, and make it a condition of tendering for government contracts that the applicant or tenderer commit to maintaining specified environmental standards up and down the supply chain[14]

- Make this a condition for the granting of regulatory flexibility (as under the US EPA's Environmental Leadership programme)

- Encourage larger enterprises to form partnerships with smaller buyers and suppliers and provide public recognition to those that do so

- Hold this out as an important feature of environmental best-practice models

- Insist on such a requirement directly in legislation

- Require such efforts to be articulated in corporate environmental reporting

Another strategy, targeted specifically at waste reduction and management, relates to the control of the suppliers of inputs. For example, it is possible for regulators to either

---

14   See e.g. OECD 1997, documenting the many ways OECD member-country governments are now using environmental criteria in purchasing.

require by regulation, or to persuade, the suppliers of various substances (against the threat of regulation in the event of refusal) to increase the amount they charge those to whom they sell their product. They might then use this amount to cover the cost of recovery and appropriate disposal. To be successful, such programmes must also include a means of encouraging or ensuring the return of unwanted substances and containers to the suppliers. This can be achieved in part by making the return of the chemicals cost-free, since the cost of recovery and disposal has already been built into the price of purchase. The most advanced example of controlling the suppliers of input has been developed under regulation in British Columbia, described in Box 2.

THE BRITISH COLUMBIA POST-CONSUMER RESIDUAL STEWARDSHIP PROGRAM Regulation of 1997 requires the producers of solvents and flammable liquids, domestic pesticides, gasoline and pharmaceuticals to take cradle-to-grave responsibility for their products that contribute to household hazardous waste. Residuals of these products are collected from householders at a network of depots. In response to the regulation, pro-ducers of solvents and flammable liquids, domestic pesticides and gasoline formed two non-profit industry associations, the Product Stewardship Group and the Consumer Product Care Association, to jointly sponsor the Consumer Product Stewardship Program.

Since the regulation prohibits a fee being charged to consumers for the return of residuals, producers through their industry associations instead instituted an 'eco-fee' to consumers at the point of purchase on new products to recover the costs of operating the household hazardous waste collection programme. These fees are collected by industry and no portion comes to government. If eco-fees are assessed, industry brand-owners must provide annual, audited financial reports to government confirming that funds were utilised for the management of post-consumer residuals.

Collected residuals are treated and contained in an environmentally acceptable man-ner by licensed waste management companies. As the programme progresses and poten-tial markets are identified, product brand-owners are expected to move up the pollution prevention hierarchy to more environmentally acceptable solutions, such as recycling, re-use and/or product reformulation.

Box 2 *Post-Consumer Residual Stewardship Program*

Source: British Columbia Ministry of Environment, Land and Parks, Environment and Resource Management Pollution Prevention Compendium, http://wlapwww.gov.bc.ca/epd/epdpa/ips/resid/index.html

Industry co-regulation and self-management

Self- and co-regulation have become increasingly popular policy instruments to curb environmental harm. We use the term 'co-regulation' to refer to a hybrid policy instrument involving a combination of government-set targets and industry-based implementation, with even the latter element being underpinned by government controls (Gunningham and Grabosky 1998: 50-56). This must be distinguished from pure self-regulation, which involves giving industry very considerable autonomy in relation to both goal-setting and implementation. Our use of the term 'co-regulation' (which we use throughout as a con-venient shorthand) resonates strongly with the concept of 'industry self-management', as used in a recent Canadian paper (Lonti and Verma 1999). This term describes the transfer of the responsibility for administering legislation and regulations from government to industry. Specifically, it involves the creation of not-for-profit, self-funded corporations, led by industry councils or similar bodies to deliver services and programmes in specific markets. Ideally, this is a policy strategy that leaves the government free to focus on its core business of setting policy directions and establishing safety standards. It also means tapping into sectoral best practices and letting industry deliver the services itself.

Almost uniformly, policy discussions about self- and co-regulation assume that these instruments can only be applied to the behaviour of *large* enterprises in particular contexts. For example, the most detailed and optimistic studies on self-regulation relate to the chemical industry's Responsible Care programme and to the US nuclear power industry's Institute of Nuclear Power Operations (INPO) (Gunningham and Rees 1997). The implicit assumption is that it is when there is a relatively small group (to minimise risks of free-riding) of large players (readily identified and with a self-interest in protecting their collective reputation) that self- or co-regulation are most credible.

In contrast, very little attention has been given to the possibility that self- or co-regulation could be credible policy options for addressing the environmental problems of SMEs. Indeed, a recent and otherwise comprehensive survey of self-regulation makes no reference whatsoever to this potential role (Priest 1998). The assumption has been that SMEs have little to gain from self-regulation, or self-management, which in this sector is likely to be a sham which will foster and encourage the 'cowboy' element (Tilley 2000; Petts 2000). Moreover, SMEs are rarely represented by industry associations with the credibility or power to enforce self-regulatory controls on their members.

Yet, as the regulatory state retreats even further, and the prospect of effective direct intervention in the affairs of myriad SMEs becomes even less plausible, there is a pressing need for other instruments to fill the policy void. We argue that, even in relation to the environmental problems of SMEs, co-regulation, though not pure self-regulation, may have a positive role to play in some contexts, and some circumstances. The crucial question is: *which ones?* Recent Canadian research suggests the following criteria for identifying a sector's readiness for co-regulation or self-management (Lonti and Verma 1999):

- A successful track record of professional development programmes by the industry, such as a code of ethics

- A documented history of consultation and partnership with government and others to solve marketplace problems

- A demonstrated capacity to perform some legislated functions on behalf of government

- The existence of a representative national or provincial industry group or association

- Proven ability to represent a balance of interests for consumer protection, including high standards of service

Even when these criteria are largely satisfied, industry co-regulation is clearly no panacea, and must be used selectively and with caution. Nevertheless, our research on the control of ozone-depleting substances (see Chapter 4) suggests it would be unwise to dismiss the potential value of co-regulation or industry self-management as a viable regulatory technique for addressing some of the environmental problems of SMEs. One is usually comparing grossly imperfect regulatory options, and, before dismissing the potential role of co-regulation as too seriously flawed to be given further consideration, one has to ask: *compared to what?* We argue that there may be some contexts in which it may represent the most viable regulatory option for SMEs, particularly where government resources are very limited.

Incentives

The potential application of economic instruments to environmental policy is well recognised (Eckersley 1995; Rehbinder 1993), largely because such instruments have the capacity to give firms much greater flexibility than command-and-control regulation in tailoring responses to their individual circumstances and achieving least-cost solutions. In *principle*, economic instruments that provide incentives should be an effective means of encouraging improvements among SMEs because 'these are the companies most likely to respond to the potential financial benefits inherent in many incentive options . . . smaller businesses appear to respond more to marginal changes in taxes' (GEMI 1999: 27). In *practice*, however, the scope for introducing such instruments in relation to SMEs may be limited.

For example, it would be impractical to design and implement a tradable permit system in relation to SMEs because of the overwhelming difficulties in monitoring and enforcing such permits when there is a large number of small, disparate polluters (or there are non-point sources of pollution). For these reasons, instruments such as tradable permits are not appropriate for curbing the environmental excesses of SMEs directly, and the main market-creation schemes, such as the US acid rain permit trading programme, engage only with large point-source polluters.

In contrast, some commentators have suggested that tax policy would probably be a considerable incentive for SMEs, and survey evidence suggests that SMEs have a clear preference for tax-based incentives over alternatives such as free energy audits (UK Round Table on Sustainable Development 1999). Tax incentives also have the attraction of being easily disseminated to SMEs via the accounting profession. Unfortunately, environmental agencies do not have control over tax policy and seeking to persuade central government treasury and key financial departments of the value of tax-based environmental incentives has usually proved to be a futile pursuit.

Moreover, even such tax incentives as have been targeted at SMEs in other countries have not been notably successful. According to a major Asia Pacific Economic Co-operation (APEC) study, 'tax incentives and grant schemes appear to be an area where numerous initiatives have been established through APEC, but with very little effect or benefit' (APEC 1999). The reasons include an inadequate level of incentives, a lack of awareness of the schemes, the difficulty of using them, and their overall complexity, coupled with a low level of interest and environmental awareness on the part of target firms.[15] It is

15 In contrast, technology transfer initiatives were considerably more successful, leading the APEC report to conclude that 'in a free market economy, informing and teaching companies about what can be done may be one of the more preferred solutions and therefore a more effective scheme' (APEC 1999).

important to note that tax incentives only work when the costs are substantially raised or lowered and that, within the framework of free inter-state and international competition, the limits to which it is possible to raise or lower costs are soon reached. Nevertheless, a range of other significant incentives capable of influencing SMEs does exist, and most of these are within the capacity of government environment agencies to introduce. In their broadest sense, these include cost reduction, customer demand, and—to a lesser extent, given its lack of reach to SMEs—regulation, all of which we explore elsewhere in this chapter.

More direct incentives can be provided through such mechanisms as audit and technology assistance and via government procurement policies. Effective dissemination of information about such instruments may be particularly important since 'previous initiatives, introduced to encourage action by SMEs have had limited success, due in part, to a lack of awareness about their existence' (UK Round Table on Sustainable Development 1999). Finally, as many economists have rightly pointed out, the removal of perverse incentives (such as fossil fuel subsidies, and particularly those that inhibit cleaner production) is probably the logical, and crucially important, starting place.

Government regulation

Consistent with a number of previous international surveys, a 2000 report by the Australian Industry Group found that 'environmental regulations were the most influential factor driving environmental management strategy, with 48% indicating regulation as a strong influence' (Information Australia 2000). This is hardly surprising, given the resource constraints experienced by most SMEs, their perception that spending on the environment represents a cost rather than a benefit, lack of awareness of win–win outcomes with short-term payoffs, and the limited incentives currently offered in most jurisdictions to engage in voluntary environmental investment.

Despite the potential importance of conventional regulation as a policy tool, its application in the case of SMEs has been undermined by difficulties of enforcement. Indeed, enforcing regulation against SMEs is a vastly different proposition from enforcement against large companies. We describe below, therefore, both a rationale for the use of government regulation and strategies to maximise enforcement, in the case of SMEs.

Even if more attention was given to the less interventionist policies for SMEs, there would still remain a role for an underpinning of regulation, both to ensure a level playing field and to forcibly change the behaviour of incompetents and recalcitrants.[16] In the absence of regulation there is commonly insufficient incentive for some firms to improve their environmental performance, while experience suggests that others still will fail to respond positively to incentives even when it is rational to do so. The importance of this is demonstrated by a review of voluntary programmes in the USA. This concluded that most have had minimal benefits either for environmental protection or for industry

---

16   On the importance of developing a mixed strategy to influence the different motivations and competences of different regulatees, see Ayres and Braithwaite 1992.

participants, mainly because of the difficulty in creating sufficiently strong incentives for industry action *without a supportive legislative and policy context* (Moffet and Bregha 1999).

Yet regulation 'in the books' is unlikely to be effective in influencing the behaviour of SMEs without credible enforcement. And credible enforcement represents an enormous challenge for already overstretched regulatory agencies, confronted with the vast numbers of SMEs. It has been suggested that, in the better resourced area of workplace safety, an SME can anticipate being inspected about once every 80 years, and the reality is that most SMEs will never encounter an environmental regulator. Indeed, in a significant minority of cases the regulator may not even know they exist!

The lack of credible enforcement is hardly likely to engender confidence in the effectiveness of environmental regulation, or to effectively deter laggards that are unwilling to take voluntary action. Unsurprisingly, a 1999 UK study found that the traditional command-and-control system is perceived by SMEs to be weak, that it has not been effectively communicated, and that it has not been credibly enforced (Petts 2000). Yet the same study also found that tough prescriptive regulation is perceived by non-management employees to be essential in bringing many SMEs into compliance.

There is no ready resolution to the problem of the vast disparity between the number of SMEs and the number of regulators, and many of the regulatory strategies we have advocated in other contexts are rendered wholly impracticable by these resource constraints.[17] But what can be done is to develop a number of strategies that at least serve to mitigate the problem: by the targeting of resources, by creating the perception that SMEs face a credible threat of inspection and of enforcement action, and by combining regulation with other, less resource-intensive strategies in such a way that only a minority of SMEs need be targeted for regulatory scrutiny.

One strategy is to reserve the threat of direct regulation for the minority that are demonstrably unwilling to take voluntary action. By passing the responsibility to SMEs to regulate themselves and devising mechanisms that, albeit crudely, alert regulators to those that manifestly do not do so, regulators can focus their enforcement resources on a small minority of the total pool of SMEs: the true laggards. To the extent that adequate statistics are available, then inspections can be targeted at those who are in the highest risk category for non-compliance. For example, a version of the Maine 200 scheme could be adopted, whereby the bottom 10% are targeted for inspection each year (Gunningham and Johnstone 1999).

Similarly, as the OECD (2000a) has argued,

> the design of compliance-friendly regulation must be supplemented by compliance-oriented monitoring and enforcement of the regulatory system. One of the most significant developments in this area is a risk analysis-based approach to targeting monitoring of compliance. Regulators are beginning to decide when and where to do inspections by analysis of data on where risks of non-compliance are likely to be highest.

However, adequate statistics are rarely available, and for the most part much cruder strategies will be necessary, such as targeting those that have failed to respond to an

---

17 See e.g. Gunningham and Johnstone 1999: Ch. 3. See also Environment Agency of England and Wales Enforcement Policy, www.environment-agency.gov.uk/aboutus/154509/154529/?lang=_e&region=, for a risk-based, targeted approach.

invitation to self-inspect and self-audit and to send the results to the agency, or whose self-reporting data is wildly at odds with that of most of their industry peers.

A tiered system of regulation may also be valuable, whereby the most substantial category of polluting industries (which may include a minority of SMEs) are required to be licensed, a middle category (which will include many SMEs) requires registration, and minor polluters are left to other mechanisms such as the use of local government health, safety and environment inspectors and licence approvals. This is the strategy adopted in Western Australia (see Fig. 1).

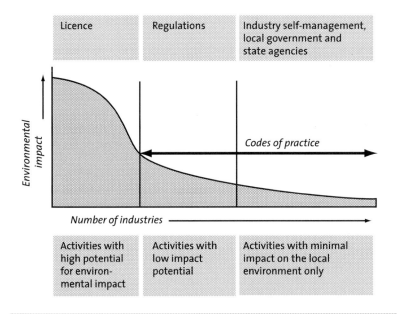

Figure 1 **Western Australia's three-tiered regulatory system**

Source: WWF 2000

It is critical to the success of this approach that a credible level of resources is devoted to the inspection and control of lower impact categories.[18] Under the Western Australian tiered approach, a reliance on public complaints is the principal detection strategy for the middle category (supplemented for some industries with occasional audits), and the main aspiration under the 'least impact' category is a reliance on local authorities to fill the regulatory vacuum. In practice, however, cracks have appeared in this strategy with insufficient external monitoring and control of some SMEs by local governments in particular. It remains to be seen whether state regulators will be successful in empowering and training often reluctant and sometimes under-resourced and unskilled local authorities to take on environmental responsibilities.

Irrespective of whether a tiered approach is adopted or not, a heavy reliance on complaints and reports from the public will remain a necessary and, given the resource constraints, inevitable part of the regulatory regime for SMEs. For example, in England and Wales it is estimated that over 80% of all environmental enforcement actions stem directly from third-party complaints or information (Petts 2000). The role of public complaints and reporting can and should be enhanced. Publicity campaigns, in particular, can be used to sensitise the public to the importance of correct environmental behaviour (e.g. the consequences of unauthorised dumping of chemicals or waste) and to encourage them to report such behaviour to the authorities. For example, 'dob in a polluter' provisions, accompanied by the relevant agency phone number and guarantees of anonymity, have proved quite successful in areas such as motor vehicle emissions. It should be acknowledged that 'after the event' complaints and reports, however, need to be accompanied by other strategies that emphasise preventing pollution at its source, and preventing accidents before they happen.

Another strategy to compensate for grossly inadequate regulatory resources is to encourage enterprises to voluntarily report their offences. For example, amendments to the Western Australian Environment Protection Act encourage voluntary reporting of offences by providing authority for the regulator to:

> Modify penalties if certain conditions are satisfied. To take advantage of these reduced penalties (for tier 2, i.e. mid-level, offences), an individual or corporation must meet the following conditions: notify the department in writing of the particulars of the occurrence allegedly constituting the offence as soon as reasonably practicable after the occurrence; take all reasonable and practicable steps to minimize and remedy any adverse environmental effects of the occurrence; cooperate with the DEP [Department of Environmental Protection, Western Australia] and provide information when requested; and take reasonable steps to prevent re-occurrence of the circumstances giving rise to the alleged offence. As has been pointed out, these provisions 'leave a person to weigh the merits of "coming clean" with a confession for a reduced penalty versus the likelihood of an offence being discovered later' (Spillman 1998).

---

18   For example, a survey of the impact of light industrial operations (including those of many SMEs) on the Swan–Canning river, Perth's major waterway, identifies substantial environmental problems caused by the absence of bunding and use of unroofed chemical storage, serious waste management problems, improper discharge to sewers and inadequate maintenance of waste-water treatment systems, inadequate storm-water treatment systems, a lack of preparedness for emergencies, and a lack of environmental awareness (Waters and Rivers Commission 1999).

Given the scarcity of regulatory resources, it is also essential for regulators to adopt enforcement measures that are both efficient (in terms of consuming the minimum of scarce resources) and effective (in terms of deterrence). The best option is often the use of administrative directions (clean-up notices, pollution abatement notices and non-statutory field directions).[19] These measures allow action to be taken swiftly without the necessity of going to court. Appeals are rare and infrequently successful. In contrast to the cumbersome and time-consuming nature of the traditional prosecution, administrative orders are a quick and simple mechanism of dealing with serious hazards immediately they are detected. Moreover, such orders are particularly flexible in that they do not necessarily specify how an employer may come into compliance, thereby leaving her or him free to choose the least costly method and avoid unnecessary expense.

On-the-spot fines also have considerable potential. The research of Gray and Scholz suggests the very considerable value even of a slight slap on the wrist, such as on-the-spot fines provide, in focusing attention on OHS issues (Gray and Scholz 1993). Research concerning on-the-spot fines in relation to OHS in New South Wales also supports this conclusion (Gunningham et al. 1998). Similarly, the broader international research on administrative penalties generally (Brown 1992) also suggests that such penalties can provide credible deterrence at a very modest administrative and legal cost.[20] Provided that use of these fines does not become a substitute for more serious enforcement action in serious or repeat cases,[21] and provided they do not serve to trivialise OHS offences through misuse, then they have considerable value.

Another strategy for maximising the impact of very limited agency resources is through rotating industry-specific campaigns and blitzes. Carefully managed, these can create the impression of a substantial regulatory capability and threat of enforcement, with only very limited regulatory resource commitment. There is convincing evidence to suggest that 'a visit [by an inspector] is likely to be relatively effective since small employers are more impressed than are many larger employers by the authority wielded by government inspectors' (Hopkins 1995), and that the very fact of an inspector's visit, *coupled with some form of enforcement action*, may have a significant impact on behaviour, even in circumstances where compliance costs will probably exceed the economic benefits to the employer of compliance (Gray and Scholz 1993).

However, given that regulators cannot afford to inspect too many premises, they may achieve considerable leverage by exploiting the gap between perception and reality. For example, they may target a particular industry sector, announce in advance that they will be inspecting premises in the area, offer a partnership approach (for example, involving self-inspections, self-reporting and subsidised audits) and reserve actual inspections and enforcement action for those who do not volunteer for this approach. Through a limited campaign, using only modest resources, they may: visit the area (making their presence highly visible); accompany this by publicity in the trade and local press; issue on-the-spot fines where they identify substantial non-compliance; hold out the threat of more serious penalties for laggards; and publicise all of this to the widest possible extent.

19   The field direction is a Western Australian innovation. The direction, which has no statutory force, is written on an official document and, according to inspectors, has a much greater impact on regulatees than informal warnings.

20   See summary and evidence cited by Industry Commission Australia (1995).

21   On the contrary, in NSW, the introduction of on-the-spot fines coincided with an increase in prosecutions.

The value of this approach for at least some types of SME is supported by the results of OHS research which suggests that the minority of SMEs that perceive themselves as a high safety, health or environment risk can be motivated by the fear that their operations might be curtailed by a regulator and/or loss of business subsequent to a high-profile accident (Wright 1998). However, such a result is only likely to be achieved if there is perceived to be 'a real possibility of detection and subsequent enforcement in the event that standards are not maintained' (Wright 1998: 44) prior to a major accident occurring.

Finally, the value of 'restorative justice' should not be underestimated. This concept implies that

> when organisations do fail to comply in the first instance, a compliance-oriented regulatory approach will attempt to *restore* compliance rather than reverting immediately to a purely punishment-oriented approach. [In the case of corporate law-breaking] the aim of restorative justice is to restore enterprises to a position where they have both the capacity and willingness to comply after they have committed a breach. It is therefore an important tool for regulators to use in responding to compliance failures (OECD 2000a: 41; emphasis original).[22]

In the environmental context, the compliance plan is perhaps the best example. This mechanism provides that, in situations where an existing firm cannot comply with new legislative requirements, it may prepare and submit such a plan. A plan can also be required where there is evidence of poor environmental performance or the presence of a high environmental risk.[23]

The contents of a compliance plan will include details of a programme of improvement aimed at achieving compliance. The plan will require the implementation of improvements in waste management and the prevention, reduction, control, rectification or clean-up of pollution or environmental harm. As such, it is both process-based (the enterprise must specify convincingly the procedures it will adopt to improve its performance) and outcome-based (at the conclusion of the plan, the enterprise will be in compliance with the legislation). Once such a plan has been approved by the relevant regulatory authority, its legal significance is that of removing any liability for prosecution for non-compliance with the specified legislative provisions. However, failure to comply with the plan will itself be an offence and the immunity from prosecution would be removed in these circumstances.

Industry and enterprise characteristics

The above analysis has evaluated a number of policy instruments which, used individually and collectively, could achieve substantially improved environmental performance by

---

22   The concept of restorative justice owes most to the work of Braithwaite (1999).
23   See further Gunningham and Sinclair 1998.

SMEs. However, it is critically important to emphasise that not all of the above instruments will be appropriate in all circumstances or to all SMEs. Effective regulatory design involves tailoring a particular combination of policy instruments to particular circumstances. Most commonly, this involves developing sector-specific policy prescriptions and recognising that, even within each sector, there are likely to be a variety of different players with different degrees of competence and different motivations.

An illustration of the advantages of taking such an approach can be seen in the US EPA's Sustainable Industry initiative. Under this programme, a sector-specific approach is adopted to identify and respond to the different motivations of different categories of firms as a necessary prerequisite to designing an optimal regulatory strategy (see Box 3).

Our own case studies of the Victorian vegetable growers, of vehicle smash repairs, and of ozone protection in the refrigeration and air-conditioning industry (Chapters 3–5), serve to emphasise the need to develop a sophisticated view of the target population of regulation, and of the political, environmental and economic context in which it operates. As we demonstrate, it will be necessary to ascertain:

- Whether or not the industry has a high profile (since this may determine the degree to which SMEs might be susceptible to public or consumer pressure)

- The level of sophistication in the industry (since this is likely to indicate its capacity to adopt complex EMSs, and the need for education and training)

- The degree of uniformity in size and management practices of an industry (the greater the diversity, the greater the need to develop different strategies for different sub-categories)

- Whether or not there is a well-organised industry association, or such an association can be established (which will be necessary but not sufficient to determine its capacity to adopt self-regulation)

- The presence of readily identifiable third parties with commercial power (raising the possibility of nurturing regulatory surrogates and establishing a supply chain approach)

- Whether the environmental issues are disparate and numerous, or focused and limited in number (in the case of the latter, regulators may achieve a bigger bang for the regulatory buck)

- The prevailing culture of the industry, and its potential receptivity to different types of approach

Conclusions

The short-term goal of regulatory policy as it relates to SMEs may be to bring the substantial numbers of laggards into compliance with environmental regulation. However, the longer-term and much more important goal should be both to reward leaders for going beyond compliance and to persuade many others that good environmental

THE METAL-FINISHING INDUSTRY INCLUDES A VARIETY OF PROCESSES THAT are performed by independent 'job shops' that tend to be small, with few available resources to address environmental concerns. The industry can be divided into distinct subgroups of firms, based on environmental performance and other characteristics (i.e. both leaders and laggards, and a number of shades in between):

- **Firms that are consistently in compliance with regulatory requirements and are most proactive in making environmental improvements to move beyond baseline compliance.** These firms tend to be more motivated by the anticipated economic payoffs of strategic environmental investments, and maintain that increased flexibility in compliance requirements would promote innovative approaches and willingness to help other firms. Barriers to proactive environmental performance by these firms may include: the presence in the marketplace of firms in the other groups against whom they must compete and a belief that inconsistency among standards and regulatory enforcement requirements between firms and at all levels of government creates competitive imbalances within industry and inhibit long-term planning, investment and beneficial risk-taking.

- **Firms whose primary environmental objective is to be in compliance with existing regulatory requirements.** These firms are driven by a strong desire to achieve and maintain compliance with government environmental requirements although they often lack the motivation and/or resources to improve beyond that level of performance. They may often lack the technical in-house expertise to develop process innovations themselves and are dependent on suppliers for information on developments in plating technology.

- **Old and outdated firms that are not sufficiently profitable to justify investments in new pollution controls and probably would like to shut down, but cannot do so because of fear of clean-up liability.** These firms may lack capital, information and shop space to improve. They do not fear enforcement because they are difficult to track down, and profit by undercutting firms in groups 1 and 2.

- **'Renegade' shops that are out of compliance, make no attempt to improve, and escape enforcement attention.** While not substantial competitors, these firms pull down the reputation of the industry and compete with the other firms by avoiding the costs of environmental investments.

Key policy options identified include the establishment of criteria for defining the environmental performance (and the development of best management practices) of firms in groups 1 and 2, with resulting eligibility for various forms of relief that would lower compliance costs; an 'amnesty' programme to encourage group 3 firms to come forward by limiting their liability in return for responsible involvement in site clean-up; and greater enforcement focus on group 4. Agencies could redirect resources by conducting less frequent inspections of responsible firms.

Box 3 *Identifying the regulatory context: the metal-finishing industry*

Source: US EPA 1994: ES 2

management and good business practice substantially coincide; that there are a range of circumstances in which using fewer resources, producing less waste and reducing pollution incidents can commonly (albeit far from invariably) lead to cost savings and competitive advantage. As Fanshawe (2000: 252) puts it:

> while the environment is still thought of as a separate issue, it will always be lower down the priorities of SME managers than, for example, finance. However, once environmental issues are accepted by SMEs as just another aspect of business efficiency, it may be said that the journey towards continuous improvement has truly begun.

Achieving these goals, and particularly the second, presents enormous challenges for seriously under-resourced regulatory agencies, and few such agencies have made substantial progress towards them.[24] Drawing from the very limited existing empirical and theoretical literature, and from our own research and case studies, we seek to summarise below the main lessons we have learned so far.

First, governments should focus initially on win–win solutions with demonstrable short-term financial payoffs. Cleaner production initiatives provide the most plausible context in which these goals are achievable. In the case of most SMEs, so little environmental initiative has been taken that much low-hanging fruit still remains to be picked, and, as a matter of enlightened self-interest, SMEs should be willing to take advantage of the considerable opportunities that exist for achieving win–win outcomes. However, a large proportion of SMEs are still unaware or unconvinced that these opportunities exist. Accordingly, educational and information initiatives, technological assistance, and a variety of related strategies described above, are capable of providing substantial policy payoffs.

Nevertheless, past experience in a number of countries suggests that many education and information initiatives have not been successful, with very poor take-up rates, despite the sometimes substantial degree of government subsidy involved. Overcoming SME indifference and incapacity are major challenges, and best results are likely to be achieved when particular attention is paid to how the message is delivered, who delivers it, how it is targeted, the value of developing partnerships with the relevant industry sector, and the use of professional third parties on whom SMEs are dependent, such as banks and accountants, to get the message across.

Second, the more SMEs can be persuaded to do for themselves, the more committed they are likely to be to the outcomes and the more successful they are likely to be in achieving them. In any event, since government inspectors only have extremely limited capacity to reach and influence SMEs directly through inspection and enforcement, few other options are available. Self-inspections and self-audits have considerable promise in this context, and, in conjunction with the use of codes of practice and checklists, can make a very considerable contribution.

Third, SMEs should also be encouraged to adopt EMSs, albeit in a cheaper, slimmed-down and simpler form than those suited to large enterprises. Such systems hold out the promise of integrating the environment with the core economic goals of the enterprise, and of building in continuous improvement and cultural change. However, the upfront

---

24   The Dutch model, albeit culturally specific, may be the most sophisticated currently available. See de Bruijn and Lulofs 2000.

costs of such systems may be considerable and the paybacks medium to long term. Accordingly, only a minority of the already better-run and far-sighted SMEs are likely to take advantage of the opportunities that EMSs provide. Nevertheless, the take-up of EMSs can be substantially expanded through the use of supply chain pressure, and large customers should be encouraged and rewarded (e.g. through awards and regulatory flexibility initiatives) for insisting that their upstream suppliers adopt an accredited EMS. Government (local, state and federal) can encourage the adoption of EMS through its procurement policies by providing a competitive advantage to firms that adopt such policies.

Fourth, the take-up rate by SMEs of cleaner production opportunities, their willingness to participate in self-inspection and self-audit, and their propensity to adopt EMSs, can all be substantially enhanced if they are provided with incentives to do so. The nature of these incentives is likely to vary depending on the particular context and is often sector-specific. No single incentive is likely to work across the board. While some of the standard economic instruments, such as tradable permits, are generally not appropriate for SMEs, a range of other incentives do hold out the promise of nudging those 'at the margin' to adopt environmental improvements, which otherwise would be insufficiently attractive to justify taking action. The literature suggests that the most attractive incentives are those that enable companies to avoid or reduce costs (Davies and Mazurek 1996). However, there is still very little empirical evidence on the effectiveness of many incentives and what evidence there is suggests that different enterprises, in different circumstances, with different cultures, are likely to respond differently to the same incentives (Stratos and Pollution Probe 2000).

Fifth, an underpinning of government regulation, coupled with (at least a perceived) credible threat of inspection and enforcement, is necessary to persuade the reluctant, the recalcitrant and the incompetent that other, less coercive, approaches are worth adopting. For example, the Minnesota experience described earlier confirms that self-inspection and self-audit take up scarce SME time and resources, and that it is only when these measures are accompanied by the threat of enforcement action for non-participation that a substantial take-up rate is achieved. Yet the numbers of SMEs so vastly overwhelm the number of inspectors that it is wholly impractical to inspect, let alone enforce, against a significant number of SMEs. Even so, the *impression* of enforcement can still be maintained through a judicious use of targeted enforcement, occasional prosecutions accompanied by broad publicity, industry blitzes, and the use of less resource-intensive instruments such as on-the-spot fines and administrative and field notices.

Finally, we have argued that, since there is no single policy option that is unproblematic and effective in isolation, the challenge is almost invariably to find an effective policy mix. A starting point in developing appropriate combinations of instruments, and in determining the sequence in which they should be used for maximum effectiveness, is to recognise the different circumstances confronting SMEs. In particular, the distinction must be made between circumstances in which they will be better off economically by improving their environmental performance (so-called win–win opportunities), and those where the converse is the case.

In the former circumstances, the emphasis should be on voluntary and incentive-based mechanisms, with enforcement only necessary for the small, irrational or incompetent minority, who remain unpersuaded to do what is in their own self-interest. In circumstances where such win–win opportunities are less evident, then a greater emphasis on incentives supported by more direct state intervention may be necessary. However, even here, formal enforcement should be seen as a last resort and, to be effective, must be complemented by the sorts of education and assistance strategy outlined above. If SMEs lack the expertise, information or technological capacity to comply, then they cannot respond either to economic incentives or to threats of sanctions. Once again, as the OECD has pointed out: 'one of the more strategic things regulators can do to increase compliance is to nurture organisational capacity to comply through education, assistance, offering consultancy services, and encouraging the growth of compliance professionals with specialist expertise in the area' (OECD 1999a: 32).

3

PARTNERSHIPS, EMSs AND THE SEARCH FOR REGULATORY SURROGATES IN THE SMASH REPAIR INDUSTRY

In this case study we explore opportunities for what are variously described as next-generation regulation, alternatives to regulation and regulatory pluralism, in one particular SME industry sector: car body shops (also known as spray-painters, panel beaters and smash repairers). In particular, we examine the limitations of traditional regulation and how they might be overcome, the potential roles of self-audit and self-inspection, environmental management systems, cleaner production and environmental partnership initiatives, and the use of regulatory surrogates such as insurance companies and other third parties.

In many respects, body shops, which repair and/or replace the external panels of vehicles that have been damaged through accidents, represent the archetypal SME 'problem' for environmental regulators. They are numerous and geographically dispersed, and many are transitory, have limited management expertise, are economically marginal, resistant to government influence, and unaware of their environmental obligations. Although individually they are not likely to be a major source of pollution, their collective impact is probably substantial. Body shops have attracted particular regulatory attention because an unduly high number of complaints are made about their environmental performance and because, overall, they have more than their fair share of environmental problems (VEPA 1995).

Body shops provide a particular challenge for regulators and regulatory policy because they do not necessarily lend themselves to the same regulatory tactics that have previously been employed on large enterprises. In this case study we identify the reasons why this is the case, we describe and evaluate new policy approaches that are currently being adopted and we consider additional opportunities for further improving their environmental performance—for rewarding leaders while dragging laggards up to a minimum legal standard. We begin by briefly describing the nature of the body shop sector and its associated environmental issues.

The Victorian body shop sector

Official statistics indicate that there are approximately 1,720 body shops (registered) in Victoria, Australia, although only about 1,200 are currently active. The vast majority are very small operations (fewer than five employees), with less than 30 businesses numbering 20–49 employees, and only one business recorded in the 50–99 employee category (ABS 2000). Approximately 800 body shops are currently registered with the Victorian Automotive Chamber of Commerce (VACC), and these, according to industry sources, tend to be 'mostly, the larger, good ones'.

A variety of activities of the body shop sector cause pollution which, in aggregate, is a serious environmental problem. Indeed, although accurate statistics are not available, the dominant view within the VEPA is that the body shop sector represents one of the worst SME pollution culprits operating within urban areas. The release of liquid wastes into storm-water and sewerage systems is an issue of particular concern. It is claimed that, even though other parts of the vehicle repair industry, such as engine and transmission repairers, produce greater overall quantities of prescribed wastes, they are far more likely to have in place appropriate waste-control systems and equipment, such as dedicated wash bays and triple grease interceptors. As one inspector commented:

> body shops don't generate any more waste than their engine repair counterparts, in fact they actually produce much less—it's just that they don't have in place effective systems for dealing with their waste.

Impressionistically, there is ready support for this view in that many body shops give the appearance of operating to lax environmental standards. For example, as VEPA inspectors point out, body shops often operate with 'shoddy' equipment and the average premises looks 'primitive'. It is common to find vehicle parts and panels strewn everywhere, vehicles being washed in the driveway, untidy and/or incomplete paperwork, oil and grease deposits on the ground and spray-painting occurring in exposed areas.[1] In more concrete terms, the activities of the body shop sector that contribute most significantly to pollution include:

- **Spray-painting.** Unlike most of the other environmental issues, spray-painting is one where the body shops dominate both in terms of overall use patterns and the potential for inappropriate discharges. Many body shops, particularly smaller ones, do not possess spray-paint booths. This increases the likelihood of problems. There is also the risk of inappropriate disposal of sludge, waste paints and thinners solvents.

- **Degreasing and parts washing.** Although body shops do not normally engage in parts degreasing, in the course of their activities degreasing of engine bays and washing of vehicle parts often occurs. In many cases, however, body shops do not have designated washing bays to carry out such activities, as they are incidental to their core activity of panel repairs.

---

1   On the other hand, a minority view, but one expressed strongly in some EPA regions, was that the body shop sector's level of pollution may be exaggerated and that other parts of the vehicle repair industry—for example, engine and transmission repairers—have a greater environmental impact, generating larger quantities of waste oil and grease overall.

- **Car washing.** Waste-water containing washing detergents may flow directly into storm-water drains. Although body shops are not a major user of washing detergents, they often do not have appropriate processes and equipment in place when they do engage in car washing.

- **Storage and disposal of waste oil.** Problems occur when oil is tipped, either deliberately or through ignorance, down storm-water drains or sewers to enter waterways, or is deposited on soil. A related problem is where waste oil, or oily or greasy rags, paper or sawdust, is burned to create polluting fumes.

- **Radiator fluid.** Individual premises are unlikely to be a major source of radiator fluid waste. However, problems may arise precisely because this lack of familiarity with proper disposal practices may result in fluid entering storm-water drains.

- **Ozone-depleting substances.** Many older vehicles contain refrigerant gas in the air-conditioning systems which, if released to the atmosphere, will contribute to the depletion of the stratospheric ozone layer.

Our interviews with a range of stakeholders (including industry respondents, industry associations, training professionals and regulatory officials) suggested a number of reasons why environmental performance in the industry was so poor.[2] Most strikingly, the large majority of body shops lacked even a rudimentary awareness of either their environmental responsibilities or the opportunities for cost-effective environmental improvement. As noted above, the body shop sector is dominated (at least numerically) by small owner-managed operations often with no more than a handful of staff. Industry respondents pointed out that they have many competing demands on their time, and that they are often operating under the added burden of financial constraints. In many, if not most, instances, according to regulators, the smaller body shops lacked even a basic understanding of their statutory obligations.

Consistent with this, a number of industry respondents pointed to ignorance and lack of skills as a major problem, with even those who wanted to 'do the right thing' often lacking the necessary expertise to translate this desire into tangible improvements. This problem was exacerbated by an aversion to paperwork and low levels of literacy. As one industry respondent stated:

> most of the people working in the industry are 'hands-on' people, with a distinct preference for practical, mechanical challenges. The prospect of wading through descriptive text, whether for the environment or anything else, is something that many, if not most, would find fairly daunting.

Most body shops also displayed little knowledge of contemporary practices capable of improving their environmental performance, or of gaining any financial advantage from

---

2 The fieldwork involved semi-structured interviews (21 in total) with a representative sample of stakeholders, including 13 enterprise managers, with the remainder including officials from industry bodies, government agencies, commercial third parties and related professional associations. This included key officials from the major industry body, VACC, and the government regulator, VEPA. Snowball sampling was used to identify individuals in the above categories most able to provide valuable insights. Interview material was supplemented by information from public documents, industry journals and reports.

doing so. For example, only a few were familiar with the concept of EMSs, and even fewer expressed a desire to adopt one. As one respondent noted: 'I have enough trouble keeping my books in order. Even if I wanted to [adopt an EMS], I wouldn't know where to begin.'

Unsurprisingly, the vast majority of body shops also displayed very little awareness of or interest in exploiting financially attractive opportunities for environmental improvement. As some of the Victorian Environmental Protection Authority (VEPA)'s cleaner production case studies demonstrate, body shops can generate savings: for example, through waste reduction (VEPA 1997). But very few had even considered such a possibility. A fairly typical response was that: 'If I have to buy a triple interceptor [a device for separating oils from water], that is going to cost me money—money that I can't really afford.'

Arguably, this reluctance of body shops to consider opportunities for cleaner production resonates with the concept of 'bounded rationality' (Simon 1992). This suggests that, even if business managers do have access to all relevant information, they may be unable to effectively absorb it and integrate its prescriptions into their day-to-day operations. This is because managers invariably focus on only a few, central core tasks. The net result is that even economically rational action to exploit win–win opportunities is commonly not taken.

However, there may be a more prosaic explanation. If the current practice is simply to dispose of waste substances illegally but very cheaply, then the notional win–win opportunities may be substantially diminished, and relate only to circumstances where recycling or re-use really is economically more advantageous than illegal disposal.

Finally, the problems identified above are exacerbated by the economic marginality of many operators. The body shop sector is undergoing a process of rationalisation. In particular, changes in the commercial relationship between the insurance industry and body shops have placed a much greater emphasis on higher volumes. This favours larger body shops, and consequently smaller players are finding it increasingly difficult to maintain their profits, and ultimately their commercial viability. This makes it less likely that they will invest in environmental improvements, even though cleaner production practices, arguably, have the potential to generate cost savings in the longer term.[3]

In any event, the pressure to both compete for, and quickly complete, jobs means that there is little time to devote to improvements in environmental management practices. The exception to this experience are the larger body shops that, through economies of scale, are able to systematically address environmental performance along with broader quality assurance issues.

---

3   Official VEPA literature makes mention of the core issue of limiting oil and grease entering waste-water streams in body shops. It is possible that the costs of triple interceptors, for example, make short-term financial benefits unlikely as the body shops are not currently being charged for this form of (admittedly illegal) form of waste disposal. Moreover, the stress and time devoted to staving off bankruptcy inevitably reduces the capacity (and enthusiasm) of management to focus attention on seemingly peripheral issues such as improving environmental performance.

The limits of traditional regulation

The Environment Protection Act 1970 (the Act) is the key environmental legislation in Victoria. This is aimed at protection and improvement of the environment through a framework of policy development and implementation mechanisms such as works approvals, licensing, improvement notices, fines and prosecution, while also providing for the promotion of pollution prevention through information and education campaigns.

Body shops are not subject to site-specific works approvals or licences under the Act. However, they must meet relevant environmental objectives set through state environmental protection policies, industrial waste management policies and technical guidelines.[4] As such, each body shop is required to conform to the minimum pollution standards contained therein. Failure to do so, by incorrectly disposing of industrial wastes, for example, can result in sanctions, including improvement notices and, as a last resort, prosecution.

In order to achieve a minimum standard of acceptable behaviour under this legislative regime, regulators, until quite recently, adopted a conventional command-and-control approach, based on enforcement of the regulatory standards and the imposition of penalties in the event of breach. The success of this approach rests substantially on the ability of inspectors to identify and penalise specific breaches to a sufficient extent to deter wrongdoing. Subjectively, it also depends on body shops *believing* that the future likelihood of such action occurring, and the anticipated severity of punishment, are sufficiently high to outweigh the advantages of continuing the proscribed behaviour. Of course, pressure and punishment will not be necessary or appropriate for all operators, and leaders (or, more broadly, better environmental performers) are not the intended targets. It is the laggards, most notably the marginal performers and the wilfully recalcitrant, who are most in need of this form of deterrence. The qualitative evidence cited above suggests that body shops may have more than their fair share of the latter group.

However, there are compelling reasons to believe that this traditional approach to regulation and its enforcement will be largely inappropriate and substantially ineffective in changing the environmental performance of SMEs generally, and of body shops in particular. Interviews with inspectors and industry respondents alike yielded a remarkably consistent picture both as to the underlying reasons why this sector has such poor standards of environmental performance and as to why command-and-control regulation is such an impoverished strategy for raising them.

A major reason cited for the limited effectiveness of traditional regulation was a lack of regulatory oversight. With at least 1,200 body shops currently active in Victoria, the chance of any one premises being visited by an inspector is very low. The problem has been exacerbated in recent years by a contraction of regulatory resources, and a redeployment of some of the remaining resources to other functions, such as information and education programmes. The remaining inspectors have their attention divided among a range of industry sectors, which limits their capacity to develop sector-specific expertise.

---

4   These include, for example: Environment Protection Act 1970; State Environment Protection Policy, The Waters of the Dandenong Valley; Industrial Waste Management Policy (Control of Ozone Depleting Substances); and Control of Wastewater from Service Station Vehicle Washes (VEPA undated).

Their effectiveness is further compromised both by the high turnover of body shops (making it difficult to build a relationship or keep track of them) and by the high turnover of VEPA staff (which contributes to a loss of 'inspectoral memory'). Unsurprisingly, the vast majority of industry respondents stated that they had rarely, or never, been visited by a VEPA inspector, and they clearly believed that the chances of breaches being detected were exceptionally slim.

A related issue was the sheer difficulty in achieving successful prosecutions if and when serious environmental breaches are suspected. According to one inspector, it is 'almost impossible' to succeed in a prosecution against a recalcitrant body shop simply because it is extremely difficult to catch them in the act. Even if, exceptionally, their illegal behaviour is detected, it must be verified in a manner likely to stand up in court, and this too is highly problematic. For example, for a complaint from a neighbour of a body shop to be verified, VEPA requires that a diary be kept, which for all but a small minority of potential complainants is 'all too difficult'. The combined effect of the difficulties of detection and of obtaining satisfactory evidence is that 'the EPA makes big noises about being tough on polluters, but in reality will not prosecute'.

The inability of the regulators to take effective action against recalcitrants and to curb free-riding in turn demotivates more responsible enterprises. As one respondent stated: 'it would make me feel a whole lot better [about] doing the right thing . . . if I knew that some of the backyard operators were getting hit hard'.

Similarly, another stated:

> I know the EPA has been informed by the public of two cases where body shops are doing the wrong thing . . . but the EPA response has been 'there is nothing we can do about it'. This makes it difficult for those body shops that have spent $50,000 [introducing cleaner technologies] if they can see that someone up the road getting away with it. No one [in the EPA] does anything.

The difficulties of pursuing a conventional regulatory approach are exacerbated by a cultural resistance to addressing environmental improvement in general, and a strong aversion to regulatory intervention in particular, displayed by many body shops. These phenomena were most pronounced among smaller, less well-resourced operations, including (but certainly not limited to) backyard operators effectively operating outside the law and, in some cases, those with links to criminal activities, notably car-stealing rackets. In the case of the latter it is extremely difficult for inspectors to conduct their activities properly, with the real threat of physical violence mentioned in several instances. As one respondent noted:

> Why is it that some body shops only seem to operate at night? We never see them open during the day. There is no doubt that stolen car recycling is going on . . . If they are prepared to flout criminal laws, what hope have we got of applying environmental laws?

According to some regulatory analysts, the problems of imposing a conventional command-and-control approach can be substantially mitigated by combining this approach with the provision of information and education and with compliance assistance (OECD 2000a). But whether government information campaigns could overcome the lack of awareness and resistance to change in the body shop sector is highly doubtful. VEPA has already engaged in a number of information and education initiatives ranging from disseminating simple promotional material to providing detailed resource-intensive demonstrations. It has developed best-practice environmental management guidelines in partnership with industry and cleaner production demonstration sites and sector-specific VEPA Bulletins which outline statutory obligations with practical interpretations for meeting these responsibilities.

However, there is evidence to suggest that body shops (and perhaps most small businesses) are resistant to external information campaigns. For example, VEPA, in co-operation with the City of Greater Dandenong, has developed an information kit targeting the motor vehicle service and repair industry (including body shops). The kit is entitled 'Is Your Business Going Down the Drain?' In addition to a number of information booklets, the kit contains a simplified information poster captioned 'Drain Strain: Think About It'. This is intended to describe good environmental practices for the industry in a simplified and humorous format. Although it is difficult to determine in quantitative terms the success of such information-based strategies, the anecdotal evidence is not encouraging. The relevant VEPA officer in the Dandenong region, for example, reported that: 'in all the site inspections I have conducted since the launch of the information kit, I have not seen one Drain Strain poster on a shop wall . . . the campaign has not been a great success.'

Certainly this example is consistent with the broader experience which demonstrates that, with limited time and resources, and a possible aversion to text-based guidelines, body shops are resistant to information campaigns, particularly those that lack follow-up face-to-face training.

To summarise, an inability to detect breaches and prosecute them successfully has a number of adverse consequences. First, it reinforces an attitude among body shops, whether they support VEPA prosecutions or not, that 'they can get away with what they like'. Second, it provides comfort to industry laggards at the expense of the industry environmental leaders attempting to make a genuine improvement. Third, it influences the allocation of resources within VEPA, generating a reluctance to put scarce resources into an area where attempts at enforcement are seen to be potentially futile. Finally, these problems are not substantially mitigated by conventional education and information campaigns which have not, to date, been markedly successful.

EPA policy response: the 'Clean Green Shop'

In recent years, VEPA, recognising the shortcomings of traditional command-and-control approaches to regulating SMEs, has initiated a very different approach: the Cleaner Production Partnerships Program (CPPP). This comprises three core elements:

- **Cleaner technologies**—promoting innovations in technologies and equipment that can not only result in improvements in environmental performance but also in commercial output

- **EMS**—introducing management processes tailored to the needs and circumstances of individual SME sectors and designed to systematically address and improve environmental performance

* **Product stewardship**—extending environmental responsibility beyond the 'factory gate' to encompass the movement of goods and services along the supply chain, and the relationships between vertically aligned commercial entities

In the case of body shops, this policy approach has translated into an initiative called the 'Clean Green Shop'. Significantly, body shops are one of the first sectors to enter into a cleaner production partnership with VEPA: first because VEPA had identified body shops as poor environmental performers and, second, because the relevant industry association, the VACC, was a willing partner in the process. In essence, the Clean Green Shop initiative entails the development of a simplified EMS, specifically designed to meet the needs and circumstances of body shops, with an emphasis on cleaner production. According to the official literature:

> The Clean Green Shop Environmental Management System enables business to protect the environment and save money. Industry has a duty of care to ensure that the environment is protected. Community expectations are also increasing. For body repairers, this Environmental Management System means reduced production costs, less waste handling and a better work environment—and that means better health and safety and a more motivated workforce. The VACC and EPA have developed a program that is tailor-made for body repairers. The Clean Green Shop is a simple, structured program that is easy to implement. After completing the program and external auditing, body repair shops receive accreditation and the right to display the Clean Green Shop logo (VACC/VEPA undated: 2).

At the heart of the Clean Green Shop is an EMS. This has three stated aims: first, to reduce the damage automotive service and repair businesses cause to the environment; second, to reduce the cost of running a business by reducing waste; and, third, to reduce the possibility of fines or other expenses that may occur as a result of mismanagement. The EMS itself consists of two components: first, a desktop 'Flipper' that contains a series of environmental standards and, second, a more detailed 'Folder'.

The Flipper is intended to be an easy-to-follow, step-by-step guide to the implementation of an EMS in a body shop. The EMS consists of 23 standards addressing discrete problems as well as broader management issues. These include, for example, environmental awareness, allocating environmental responsibilities, recycling, cleaning up spills, conserving energy, legal compliance, and reviewing performance. Guidance is given on how to implement each standard, beginning with the appointment of an 'Environmental Representative' to oversee the process. Each standard is intended to be addressed individually with the participation of all staff to determine appropriate actions. The Environmental Representative is required to 'authorise each Standard by signing and dating the Action' on the Flipper. Once this has been completed, all staff are expected to comply with the agreed methodology, and the process is repeated for each of the other environmental standards.

The Folder is much more detailed than the Flipper, and is designed to be used mainly as a periodic reference and as the repository of EMS record-keeping. The Folder consists of five parts. First, there is an 'EMS Manual' containing specific environmental procedures which, collectively, constitute the substance of the EMS, and which must be formally implemented. Second are 'Good Practice Guidelines' which are intended to provide practical advice on how body repair shops should go about meeting their statutory obligations, as well as detailing how others have achieved and benefited financially from improved environmental performance. Third is an 'EMS Checklist' consisting of a series of photocopy masters that are used to identify current environmental performance and progress towards the implementation of the EMS environmental standards. Fourth are 'Position Descriptions': another series of photocopy masters which, in this case, identify the roles and responsibilities of each body repair shop staff member, as well as outlining their prospective training programmes. Fifth are 'Forms and Reports', which provide the necessary record-keeping component of the EMS in order to demonstrate effective implementation.

Some features of the Clean Green Shop EMS are quite distinct from those of mainstream management systems such as ISO 14001 and EMAS. For example, instead of striving to achieve best-practice environmental management, the concept of good practice was introduced in order to emphasise realistic and achievable improvements in environmental management that reflected the resources, cultural attitudes and technical sophistication of the sector, and which were geared to at least achieving compliance. Again, given the aversion of many body shops to detailed paperwork and record-keeping, the emphasis is on environmental management procedures and guidelines that are not only simple to interpret and understand but also easily monitored and recorded. Efforts have also been made to keep the costs as low as possible, and the fact that the Clean Green Shop EMS can be purchased 'off the shelf' has achieved significant economies of scale.

Other features of the Clean Green Shop EMS are more traditional. For example, the basic tenet of cleaner production—that environmental improvement occurs in tandem with economic benefits—is central to the strategy. This point was deemed to be essential in order to successfully market the programme to body shops. Indeed, according to the VACC: 'our insistence on commercial benefits in conjunction with environmental improvement was a non-negotiable condition of the VACC's participation in the entire program'. Also conventional is the requirement that all participating body shops submit to an external audit (conducted by a VEPA/VACC-approved auditor) to determine the existing level of compliance with statutory obligations, to identify opportunities for joint economic and environmental improvement, and to provide for annual reviews of progress. The use of external auditors was deemed by both VEPA and the VACC to be essential to providing the necessary credibility and accountability to the CPPP initiative.

Finally, the Clean Green Shop initiative also has the potential to act as an industry-wide environmental code of practice. Thus even body repair shops choosing not to submit to formal external audits still have the opportunity to follow the Clean Green Shop good practice guidelines independently. In so doing, they can position themselves both to meet and exceed regulatory requirements and to take opportunities to achieve win–win outcomes suggested by the guidelines.

Policy implications

The Clean Green Shop EMS initiative is an innovative addition to the regulatory landscape, although, given its early stage of implementation, it would be premature to make a judgement as to its success in its present form. However, based on the regulatory experience of body shops to date, it is possible to identify a number of challenges that will have to be overcome if this initiative is to be successfully and widely adopted and if the policy potential of this development is to be fully realised.

A key issue for policy-makers is extending the adoption of the Clean Green Shop beyond a relatively small number of larger and better-resourced body shops (the industry leaders) to the smaller, more economically marginal body shops. These include those with limited management expertise, those resistant to intervention, backyard operators and those with an aversion to written documentation. If such body shops are not targeted, there is a danger that the initiative will simply be 'preaching to the converted'. That is, it may be directed only to the more sophisticated body shops that are already familiar with (and/or motivated to find out more about) the concepts of quality assurance and EMS, and which are both more capable and more motivated to implement them.[5]

This raises a potential dilemma for policy-makers. On the one hand, it is desirable to attract industry leaders to the initiative to ensure its initial success. On the other hand, there is the danger that it will never extend beyond this group to mediocre performers and, ultimately, to the industry laggards. It would be unfortunate if those most in need of the benefits that an EMS approach might provide were to be the least represented category in the programme. Yet this is a serious possibility. The new initiative was developed jointly by VEPA and the VACC, which represents 800 of the 1,200 or so active body shops in Victoria, and within this group it was the largest operators that, by all accounts, had the greatest voice. It is not inconceivable, therefore, that the worst environmental per-formers in the sector exist outside the VACC membership. If this is the case, there is a possibility that the very prominence of the VACC in the Clean Green Shop initiative may inadvertently deter non-members from participating.

Another potential disincentive to the smaller, least sophisticated body shops is their aversion to text-based instructions and comprehensive record-keeping. VEPA and the VACC have been sensitive to this issue and, as a result, have attempted to develop an EMS with easy-to-understand, self-explanatory instructions and simple record-keeping require-ments. It also anticipated that the external audits will provide some training in the adoption of the EMS. Nevertheless, even these steps may prove insufficient to overcome some of the resistance of smaller body shops that lack resources and management exper-tise, and may have rudimentary literacy skills.

The requirement for an external audit may also dissuade some from participating in this initiative. There was general agreement during its development that annual external audits would provide an important objective and independent assessment of environ-mental management performance, that this would have beneficial flow-on effects to other areas of business management, and that such audits provided the necessary credi-bility and independence to the initiative. Nevertheless, there were concerns that submit-ting to an external audit might expose inadvertent environmental breaches, thus leaving a body shop at risk of punitive VEPA sanctions. Notwithstanding VEPA willingness to exercise discretion on this issue, such concerns remain, particularly among the smallest enterprises.

---

5   Many of the participants in the Clean Green Shop pilot phase may be described as environ-mental leaders within the body shop sector, who had commonly already instituted significant environmental improvements prior to their participation in the pilot phase.

Unfortunately, these disincentives to participation are unlikely to be counterbalanced or overcome by the current incentives being offered under the programme. For example, the ability of participants to display the Clean Green Shop sign outside their premises is of relatively little benefit since the bulk of their business does not come directly from the public, but rather from the insurance industry. In any event, members of the public have not to date shown particular environmental sensitivity to the behaviour of body shops, nor have they sought to place any special environmental demands on the industry.

Another incentive is the prospect of a financial return from the adoption of cleaner production practices. However, many of the smaller, less sophisticated body shops tend to believe that the environment is a cost rather than an opportunity and this perception is deeply entrenched. As indicated above, they may be substantially correct in this assumption if their current practice is to dispose of waste products illegally (at minimal cost), leaving only limited savings to be made from re-use and recycling.

Finally, the Clean Green Shop offers the incentive of subsidised environmental audits. Leaving aside the question of whether laggard body shops would welcome an external audit, the costs (even with a subsidy), although seemingly low, may still be sizeable enough to dissuade smaller, economically marginal shops from participating. If a substantially greater proportion of body shops are to be persuaded to adopt the EMS route under the Clean Green Shop initiative, then some strategy, well beyond the scope of these conventional incentives, will be necessary.

This critique suggests that there are two outstanding (and related) policy issues for environmental improvement in the body shop sector: how to increase the take-up of the Clean Green Shop initiative; and how to deal with those who either fail to take it up, or fail to do so effectively.

Expanding the reach of the Clean Green Shop initiative: the use of regulatory surrogates

Confronted with the limitations of conventional regulation, and with no prospect of increased regulatory resources on the horizon, policy-makers must cast their net more widely in the search for viable instruments and parties capable of achieving their environmental policy goals. The two main options are either self-regulation or de facto regulation by third parties, otherwise known as regulatory surrogates. We return in the next section to the potential role of self-inspection and self-audit as instruments arguably capable of reinforcing direct government regulation. But, first, we examine the crucially important role, at least in this industry, of regulatory surrogates (also known as third-party policing, third-party liability and third-party enforcement) (Anleu et al. 2000).

Regulatory surrogates are entities other than government regulatory agencies that are in a position to exert pressure over enterprises (in this case, body shops) such that they improve their environmental performance at least to the level required by regulation. They have the potential to either complement or replace the activities of traditional regulators. Our argument will be that, if the VEPA is to extend its influence beyond its own very limited resources, and if the Clean Green Shop initiative is to be expanded beyond industry leaders, a crucially important task will be to harness the power of such regulatory surrogates as instruments of informal social control capable of shaping future cleaner production outcomes.

Three such regulatory surrogates potentially exist, capable in theory of greatly enhancing the uptake of the Clean Green Shop EMS through the application of external pressure. The first is the VACC, which has so far relied on its powers of persuasion to convince its body shop membership base to participate. In this regard, the VACC has an educative role which, in combination with its position of industry authority, may be effective in expanding body shop participation beyond the pilot phase. It also has the capacity, if this strategy proves to be ineffective, to make the implementation of a Clean Green Shop EMS a compulsory requirement of VACC membership. Indeed, at the time of writing, this option seems likely. However, if the VACC acts alone, then those members who are reluctant to take up the Clean Green Shop initiative may weigh the costs of participation against the costs of withdrawing from the VACC. Unless there are substantial disincentives to do so, a significant number of members may take this option.[6] And, in any event, this VACC initiative will have no direct influence on non-members, many of whom have great environmental problems.

A second credible source of external pressure to participate in the Clean Green Shop EMS programme is the vehicle insurance industry. This source of pressure is potentially extremely effective, because the body shop industry is heavily dependent on the insurers, relying on them for up to 80% of its total work. This dependence gives rise to a substantial imbalance of commercial power, and gives the insurers the capacity to require the body shops to adopt specified environmental practices. But will the insurance industry be willing to adopt that role? As we will argue, the circumstances are particularly ripe for it to do so.

There has been a long history of antagonism and antipathy between the body shops and the insurance industry, caused substantially by allegedly corrupt and unprofessional behaviour on the part of the former. In particular, it is widely believed that some individual body shops systematically colluded with their peers to artificially boost prices, under the system whereby three quotes were needed before approval would be given by the insurance company for a damaged vehicle to be repaired. This situation prompted the insurance companies to seek to exert much greater control over the body shops. First, they have sought to direct the work to preferred body shops, in particular those they consider to be reliable both in terms of quoting and completing the work to a consistently high standard. Second, they have redesigned insurance policies so the insurer controls to a very high degree the quoting and repairing process by dealing directly with insurance holders, referring them to preferred body shops with pre-set prices.

The move away from the 'three quotes' system to one in which insurers assume a central role in determining pricing has had a fundamental impact on the nature of the relationship between insurers and the body shops. Suspicion and antagonism are being replaced (at least in part) by an atmosphere of co-operation that never before existed.[7] This move has also produced greater efficiencies in the body shops (with less time wasted

6   The VACC has already provided one clear incentive for member participation: it has 'endorsed contractors for waste disposal and other services who are prepared to offer special rates and arrangements to members' (VACC/VEPA undated: 4).

7   Insurers and body shops negotiate in advance agreed pricing structures to accommodate the different types of panel repair procedure. However, the insurers will accept some variation in job prices, as long as the reasons for such variations are documented.

on quoting, and more time and resources devoted to completing the work), and greater professionalism as the focus shifts from pricing to process and outcomes. But how does the insurer know that the body shop is acting in good faith? The body shop is required to complete its work in accordance with documented guidelines, and written records must be kept to reflect this. In short, body shops have been forced to introduce a system of quality assurance.

In the future, the economic stability and prosperity of individual body shops will be intimately tied to the strength of their relationships with the major vehicle insurers, and demonstrating quality assurance will be a central component of maintaining such relationships. Against this backdrop, it would be a relatively short step to extend the current Quality Assurance approach (a system that is arguably a natural precursor to the use of EMS) to one of requiring environmental quality control by virtue of the Clean Green Shop EMS. But will the insurance industry be willing to do so?

The short answer is *probably*, so long as it is either (a) the body shops or (b) vehicle owners, not the insurance industry, that bears the costs. The former outcome is the more likely one, in view of the market power of the insurance industry. Body shops will be unable to pass on the costs to the insurers because the latter wield far greater financial muscle and have the capacity to impose their will on the body shop community through a strategy of 'divide and conquer'. If one body shop attempted to pass on EMS costs, the insurers could simply redirect their business elsewhere. As a result, body shops would be left with no choice but to absorb such costs.

The latter outcome can be achieved if the insurance industry as a whole agrees to subsidise the costs of EMS implementation in the body shop industry, while passing on the extra costs to vehicle owners. This option would serve to overcome both the resentment of body shops in being forced to participate in and bear the additional costs of an EMS adoption programme, and the commercial concern of insurers not to suffer a pricing disadvantage in comparison with their competitors. To this end, it has been suggested by the VACC that a levy of A$10 on each car plus an additional A$1 per thousand cars be imposed voluntarily on the insurance industry to subsidise body shops participating in the Clean Green Shop EMS programme. Although this levy approach might appear attractive, particularly from the perspective of the body shop sector, it remains to be seen how it might be initiated and instituted, or whether an individual insurance company might defect from this arrangement.

The essential point, however, is that, under either of these options, there is no economic disincentive to the insurance industry in adopting this surrogate regulatory role, and in this context it may be much more amenable to doing so. There is already a model for this approach in the form of the arrangement between the Hamilton District Autobody Repair Association in Ontario, Canada, and the insurance industry (AIIIC 2000), whereby access to insurance-covered auto business is conditional on honouring the industry association's environmental code of conduct.

In Victoria, it may not be too difficult for the government to prevail on the insurance industry to do similarly (and, in principle, NGO pressure might have a similar effect).[8] Indeed, discussions have already taken place between the VACC, VEPA and insurers to explore the possibility of preferential commercial treatment of Clean Green Shops by the

---

8 The insurance industry is experiencing increasing criticism from environmental groups. See e.g. FOE 2000. Given its vulnerability, it might well acquiesce to an initiative that would be largely cost-free to the industry.

insurance industry, and the insurance industry has acknowledged that good environmental management 'will be one of the things that [it] will support at the policy level' (*Environmental Manager* 2000). The Royal Automobile Club of Victoria had also indicated an intention to proceed down precisely this path, but its recent merger with the New South Wales Royal Motoring Association has put this development on hold at least temporarily. Other recent mergers and acquisitions in the insurance industry have made it difficult to develop a coherent and consistent insurance industry policy stance. Nevertheless, the long-term policy direction, and its potential power to influence the environmental practices of body shops, is clear.

A third possible candidate for the role of regulatory surrogate is the chemical industry, in particular the manufacturers of automobile paints. Body shops are dependent commercially on the supply of paint for spray-painting, and this provides an opportunity for environmental product stewardship via supply chain pressure in relation to this aspect of their activities. Specifically, chemical companies, by way of their commercial dominance, could require that their paints are supplied only to those body shops that have in place a Clean Green Shop EMS.

There are two reasons why such chemical companies might engage in this type of activity. First, the chemical industry in Australia is already covered by the Responsible Care initiative. As described elsewhere, this is an ambitious self-regulatory programme which already contains provision for environmental product stewardship (see e.g. Gunningham and Rees 1997; Gunningham and Grabosky 1998; King and Lenox 2000; Metzenbaum 2000). Second, to the extent that chemical companies operate in Victoria, VEPA may be able to facilitate the application of supply chain pressure, by insisting on it as a term of the regulatory licence. For example, some chemical companies are already covered by VEPA's accredited licensing programme (see Chapter 8), which provides them with considerable regulatory flexibility, and it would be a natural extension to such a programme, as indeed it has been under some of the US EPA's Reinventing Environmental Regulation initiatives (Davies and Mazurek 1996), to insist on such a condition. Again, this would seem to be an ideal vehicle to incorporate the preferential treatment of Clean Green Shops by large chemical companies.

Some chemical companies are already beginning to implement precisely this approach to product stewardship. For example, the Dulux company has stated that 'developing better relationships with its customers ensuring ongoing service and product purchase [and] sustainability of its customers is key to its survival', and for this reason has 'embraced the concept of product stewardship and is assisting its customers to survive in an environment of increasing waste disposal costs and community expectations' (Hamilton *et al.* 1999: 5). In pursuing this initiative, Dulux (ICI Autocolour) has received assistance from the CPPP to develop and implement environmental management systems for three of its customers in the vehicle repair industry, one of whom has now had its EMS certified to the ISO 14001 standard.

A regulatory safety net

For those body shops that are unwilling, despite the external pressures and the influence of the insurance industry or the paint suppliers, to adopt the EMS approach, or who adopt it only in tokenistic form (as exposed through third-party audit), some regulatory under-pinning will be necessary to ensure that minimum environmental standards are achieved and to ensure a level playing field. This will also be necessary as regards the 20% of body shop work that comes directly from vehicle owners, where there is unlikely to be any coherent commercial pressure for environmental improvement.[9]

But, as we have indicated, the options available to regulators are very limited. The large number of body shops, the small number of regulators and the difficulties of establish-ing breach of regulation are among the principal reasons for this. If owner-managers have virtually no chance of being inspected, and even less of being punished for wrong-doing, then environmental management will not be prominent in their thinking. Com-pounding this problem has been the policy decision of the VEPA to channel resources away from inspections to more innovative approaches such as information and educa-tion, and cleaner production partnerships.

There is no ready resolution to this problem, but it can at least be mitigated. In partic-ular, an underpinning of government regulation coupled with (at least a perceived) credible threat of inspection and enforcement is necessary to persuade the reluctant, the recalcitrant and the incompetent that other, less coercive, approaches are worth adopt-ing. For example, the EMS Flipper could be used as the basis for enterprises to self-audit, with the results being reported to VEPA. The best results are likely to be achieved if all body shops are required to register or obtain a licence (which enables ready identification of all premises and facilitates ready closure of those that fail to meet the requirements), and if letters are then sent to all owners notifying them of the audit programme but also threatening inspection for those who did not participate in the programme.

There is empirical evidence to suggest that self-inspection and self-audit take up scarce SME time and resources, and that it is only when these measures are accompanied by the threat of enforcement action for non-participation that a substantial take-up rate is achieved (NAPA 1997). Yet the numbers of SMEs so vastly overwhelm the number of inspectors that it is wholly impractical to inspect, let alone enforce, against a significant number of SMEs. Even so, the *impression* of enforcement can still be maintained through a judicious use of targeted enforcement, occasional prosecutions accompanied by broad publicity, industry blitzes, and the use of less resource-intensive instruments such as on-the-spot fines and administrative and field notices.

One option for at least partially compensating for the paucity of government inspectors would be to combine the inspectoral resources and/or functions of related regulatory agencies (the most obvious combination being safety, health and environment). In par-ticular, OHS inspectors could be given authority to identify a basic set of environmental breaches, and call in VEPA officials if warranted, and vice versa. Certainly, there is suffi-cient overlap between OHS performance and environmental management performance to make such an arrangement viable. There have already been some attempts in Victoria to institute such an arrangement with various agencies implementing memoranda of understanding, but none has so far been pursued with any political or administrative conviction.

---

9   Some vehicle owners choose to avoid insurance claims for minor repairs in order to avoid the loss of their no-claim bonus. Others are simply 'doing up' older or high-performance or 'street' vehicles.

Finally, the industry and some areas of government, notably the Minister for Small Business, are considering the possibility of introducing a licensing scheme for the industry. The prospect of industry licensing opens up a number of possibilities for pursing environmental improvement, in particular through self-regulatory provisions. Licensing conditions could, for example, require all prospective licensees to outline how they proposed to achieve regulatory compliance and to require a periodic self-audit as a condition for licence renewal. A significant advantage of such a self-regulatory licensing arrangement would be the capacity to reach those smaller body shops that have to date largely eluded regulatory enforcement provisions. A significant disadvantage is that it would also increase the demands on scarce regulatory resources and run counter to the trend of only requiring licensing of major polluters in order to make most efficient use of those resources.

Exiting the industry

At the 'bottom of the pile' in the body shop industry are the true laggards: a group of players who are both economically marginal and environmentally irresponsible.[10] Some of these allegedly supplement their legitimate income by criminal practices relating to the handling of stolen vehicles, which are resprayed for re-sale. In the United States, where more substantial regulatory resources exist, attempts have been made to 'come down heavily' on this group. There have also been proposals (as in the case of metal finishers) to facilitate its leaving the industry: for example, by waiving potential liability in relation to contaminated sites (US EPA 1994).

In Australia, such options are not available. However, the recent move by the insurance companies towards fixed-price repairing may indirectly achieve a similar result. Fixed-price repairing has produced a polarisation in the body shop industry between larger, more sophisticated and successful premises and their smaller, less sophisticated and more economically marginal cousins. As we noted above, there are approximately 1,700 body shops in Victoria. Of these, about 800 are accredited with the VACC. However, there are only about 350 'mainstream' body shops currently in operation.[11] It is these 350 industry 'leaders' that are the major beneficiaries within the body shop industry of the move towards quality assurance systems, and it is those with the least resources and sophistication who are 'left in the cold'. The initiatives of the private sector may yet achieve what government regulation has so far failed to do, in terms of removing the very worst firms from the industry entirely.

10 It is widely accepted within industry circles that there are presently too many body shops for the amount of work available. Many are probably too small to be viable. The sector is expected to undergo a period of rationalisation in coming years. Indeed, the VACC is 'encouraging smaller operations to close down', albeit with limited success. The VACC anticipates that the bottom 5% of the business will cease trading in the next few years.

11 These may be defined as having more employees, usually 12–15, and a workshop in the order of 8,000 ft^2 (as a minimum), and being located in regional centres with their own work 'catch–ment'.

Conclusion

Until recently, many regulatory authorities seeking to improve the environmental performance of SMEs relied substantially on the use of traditional command-and-control regulation (albeit supplemented to some limited extent by 'outreach' activities), despite the manifest lack of capacity to carry out this policy successfully. Over time, a growing number of policy-makers have come to recognise that very different strategies will be necessary to address the challenges posed by SMEs. This has certainly been the case in Victoria, where the Environment Protection Authority has invoked an alternative approach which, while in many respects context- and industry-specific, may nevertheless provide broader insights into the potential value of environmental partnerships, EMSs, and regulatory surrogates as applied to SMEs. The use of cleaner production partnerships has been one innovative strategy to emerge out of this regulatory re-evaluation. Cleaner production presents particular difficulties for the SME sector. The reasons for this include a lack of information and expertise, problems with cash flow, lack of suitable and available technologies with a short payback period and a tendency for management to focus exclusively on core business decision-making processes. Nevertheless, the joint VACC/VEPA Clean Green Shop initiative highlights that an EMS has the potential to bring the benefits of cleaner production to an SME sector with a reputation for poor environmental performance. However, it is only in limited circumstances that such an initiative has the potential for success. These include:

- A supportive industry body, such as the VACC, that is willing not only to develop an EMS but to promote its use among its membership

- An EMS that is not just generic but rather is tailored to the particular needs and circumstances of the small business sector in question, and which is simple, readily accessible and affordable

- A focus on practical and achievable environmental improvements to bring a poor-performing small business sector 'up to scratch' rather than aiming for unrealistic, best-practice environmental management standards (at least initially)

- Use of external auditors to identify opportunities for cost savings through environmental improvement, to provide practical guidance on implementation, to conduct annual audits that act as a spur for continuous environmental improvement and, in so doing, provide overall credibility of the EMS programme

- The existence of considerable win–win opportunities, since, in the absence of these, there may be insufficient incentive for enterprises to participate in cleaner production initiatives[12]

But even this will not be enough to touch more than a very modest proportion of industry leaders, and may leave untouched those most in need of improved environmental performance.

---

12 On the complexities of determining whether such opportunities really do exist in a particular context, see Reinhardt 2000.

Arguably, the single most potent strategy for improving the environmental performance of the large majority of industry enterprises, and for ensuring that they adhere to a systematic approach to environmental management, will be to harness the use of regulatory surrogates. The insurance industry in particular, by virtue of its commercial power and the heavy dependence of the large majority of body shops on it, could play a crucial role in improving environmental performance. It could do so by giving its business only to body shops that meet specified environmental criteria. With such a strong incentive structure (which at the time of writing seemed likely to develop), it should be possible to support a tiered approach whereby firms progress from an initial self-inspection and self-audit to a simplified EMS and third-party audit in the longer term. This approach may also have the side-effect of marginalising and ultimately excluding the worst performers.

Finally, even the very best instruments are unlikely to work optimally in isolation, and an underpinning of government regulation, harnessing bluff and the strategic use of extremely limited regulatory resources may be necessary to ensure that industry laggards do not completely escape their environmental responsibilities.

REFRIGERATION AND AIR-CONDITIONING
Co-regulation and industry self-management

As indicated in Chapter 2, self- and co-regulation have become increasingly popular policy instruments to curb environmental harm in respect of large enterprises. In this case study we focus on how co-regulation has been used to address the control of ozone-depleting substances in Western Australia, a problem that concerns SMEs almost exclusively. We examine the environmental and policy context in which ozone co-regulation evolved, we evaluate the outcome and we explain its relative success, despite the fact that it had almost none of the classic attributes identified in the literature as necessary for such success. We also identify the shortcomings of the current regime and how these might be overcome. Finally, and more broadly, we seek to identify circumstances in which co-regulation and self-management are likely to represent a credible (and sometimes perhaps the only credible) policy option for controlling the environmental behaviour of other types of SME.

The evolution of ozone (co)-regulation

In the 1980s, there was widespread scientific and community concern at the so-called 'ozone hole' that had appeared over the Antarctic region in the Southern Hemisphere and, to a lesser extent, the Arctic region in the Northern Hemisphere. Scientific research at the time concluded that the decline in the stratospheric ozone layer could be attributed directly to the increased presence of halogenated chemical compounds in the upper atmosphere which reacted with and broke down ozone. The source of these compounds was identified as a range of halogenated hydrocarbons, notably chlorofluorocarbons (CFCs), chlorinated solvents and halons, used in a variety of industrial and domestic applications, including aerosol propellants, refrigerants and flame-retardants.

After some initial scepticism, the overwhelming evidence provided by scientific researchers convinced governments throughout the industrialised world that substantial policy action was required. This led to extensive international negotiations that culminated in the creation of the Montreal Protocol on Substances that Deplete the Ozone Layer (the Protocol) in the late 1980s. The purpose of the Protocol was to phase out the use of ozone-depleting substances, and it required signatories to implement progressive timetables for achieving this within their own jurisdictions. Australia became a signatory to the Protocol in 1989, along with 27 other countries.

The phasing-out of ozone-depleting substances in Australia necessarily required a co-ordinated policy response on the part of Commonwealth, State and Territory governments. A demarcation of roles and responsibilities was established under the National Strategy for Ozone Protection. This involves the Commonwealth controlling the import, export and manufacture of ozone-depleting substances, and products containing such substances (through the Commonwealth Ozone Protection Act 1989 and subsequent amendments), and the states and territories addressing their sale and use. One outcome of this differentiated policy response is that the states and territories have tended to implement industry-specific controls, whereas the Commonwealth has adopted across-the-board supply controls that do not discriminate between different industry sectors.

The Western Australian government responded promptly to the need for state-based end-use controls to complement national legislation, and was the first Australian state to introduce specific ozone-protection legislation (the Environmental Protection [Ozone-depleting Substance] Policy 1989). This legislation was subsequently superseded by the Environmental Protection (Ozone Protection) Policy 1993.

What characterised the Western Australian response was the gradual emergence of, and increasing reliance on, industry co-regulation as its central means of achieving ozone protection across the three industry sectors that were major emitters of ozone-depleting substances: automobile air-conditioning, commercial building flame-retardation, and refrigeration and air-conditioning.[1] This approach placed the relevant industry association in a crucial co-ordinating role, with little, if any, day-to-day to management kept in the hands of the government environmental regulatory authority. This strategy contrasts sharply with conventional regulatory measures that rely substantially on the external enforcement of mandatory controls by government inspectors.

It is this unusual reliance on co-regulation (for an SME sector) that is the principal concern of this case study. Our particular focus is on refrigeration and air-conditioning because, of the three sectors targeted by the regulation, this was the most challenging to self-regulate effectively, because it was not serviced by any existing and representative industry association. As we will see, the relative success of this approach, despite the necessity to manufacture the creation of such an association, suggests that the conventional dismissal of co-regulation as unsuited to SMEs is premature.

---

1   By this time the use of ozone-depleting substances in various pressurised personal hygiene and domestic products had been voluntarily discontinued.

The Western Australian response

Western Australia introduced the Environmental Protection (Ozone-depleting Substance) Policy 1989 to control the release of CFC (and other ozone-depleting) gases to implement recommendations of the National Strategy through ambitious end-use controls (EPA 1993). The initial Western Australian response provoked concern from eight trade institutes and associates, because it threatened the autonomy of industry and imposed additional, and in their view unnecessary, costs. There was a consensus, however, that the industry would not be able to mount an effective and collective challenge to the policy because it was not effectively organised to do so. To this end, the industry-based Refrigeration and Air Conditioning Registration Board (RACIRB) was formed in 1992.

Shortly thereafter, when the Western Australian government sought to revise and improve its ozone-protection policies in consultation with a wider number of stakeholders, RACIRB was well placed to argue its case and was largely successful in doing so. Together with the Motor Trades Association, RACIRB negotiated successfully for the introduction of a co-regulatory regime to control the use of ozone-depleting substances. This served to keep it 'largely in control of its own destiny', and had the effect of substantially easing the administrative burden on the industry. The outcome of this negotiation process was the Environmental Protection (Ozone Protection) Plan 1993.

During these negotiations the Department of Environmental Protection, Western Australia (DEP) was more of a willing partner than an adversary. As we will see, DEP believed there to be considerable advantages in the co-regulatory approach in terms of making best use of its scarce regulatory resources. Whether this approach also served the broader public and environmental interest is a crucial question that we address in the following sections.

The co-regulatory regime

This new legislation had a number of important objectives: it repealed existing (and now outdated) ozone-protection legislation; it applied to a much larger array of ozone-depleting substances, including halons and hydrochlorofluorocarbons; and it brought all regulatory responsibilities under one statutory umbrella. Significantly, at least from the perspective of this case study, it also provided the legislative mechanism that allowed DEP to devolve key ozone-regulatory responsibilities to designated industry associations. In this regard, the relevant part of the legislation is Part 4, including, in particular, the following sections:

> 39. (1) The chief executive officer may from time to time approve of a person, body or association as an issuing body for particular types of authorisations that may be granted for the purposes of this policy.
>
> (2) The chief executive officer may alter or revoke any approval given under subclause (1) at any time by notice in writing given to the person, body or association concerned.

40. An issuing body is empowered to grant or cancel authorizations in accordance with this policy, and to receive and retain fees payable under this Part in respect of those authorizations.

Although not specifically referred to in the Western Australian ozone-protection legislation, a number of additional policy implications flowed from the devolution of responsibility for the issuing of authorisations to industry associations: in this particular case, RACIRB. Collectively, the measures that we describe immediately below make up the co-regulatory programme of the refrigeration and air-conditioning industry sector.

As noted above, RACIRB was formed in 1992 with a membership that drew from individuals represented on at least one of the recognised institutes or associations active in the industry. RACIRB is funded entirely from industry fees generated by registration and accreditation. It is a non-profit organisation, and all its members act in a purely voluntary capacity. RACIRB currently has a membership of 12. It operates on a 'shoestring' budget with only one staff member, the full-time Executive Officer (who operates from a home office and carries out *inter alia* administrative and inspectoral duties but is not a member of RACIRB). The Department of Environmental Protection is not a formal member of RACIRB either, but does enjoy official observer status. Apart from the departmental representative, there is no outside membership of RACIRB (for example, an environmental organisation).

The initial purpose of RACIRB was to seek to modify the prevailing ozone-protection legislation (that is, the Environmental Protection [Ozone Protection] Policy 1989) such that industry concerns, specifically the desire for greater industry self-regulation, were incorporated into replacement legislation. This, obviously, was a function with a limited life-span. With the advent of replacement legislation (that is, the Environmental Protection [Ozone Protection] Policy 1993) that largely met industry preferences, RACIRB assumed a new role. It became the 'designated issuing body' that not only administered the registration and accreditation process but also took on enforcement and education and training responsibilities. We deal with each of these three functions more fully below.

Registration and accreditation

The major obligation of companies and individuals operating in the industry in Western Australia (specifically, those involved in the 'design, manufacture, installation, service or decommissioning' of equipment containing or likely to contain ozone-depleting substances) is to obtain registration and accreditation, respectively, before they can operate or obtain work in the industry, or indeed purchase replacement refrigerant gases.[2] In order to receive registration, a company or business must: pay a very modest annual fee;

2   A number of industry respondents advocated introducing a licensing system for the industry. This would place the industry on a par with other licensed trades: in particular, electricians and plumbers (in NSW, for example, the trade is fully licensed). It was claimed that licensing would remove those that are most likely to conduct substandard work, including substandard environmental practices. Some non-industry respondents were opposed to the licensing on the grounds that it is incidental to environmental performance. Others were suspicious, or at least sceptical, of the motives of those seeking to pursue licensing under the guise of environmental protection, and suggested that it may be more about industry protection.

employ only accredited staff (see below); and possess (or have access to) ozone-depleting substances reclamation equipment. In order to receive accreditation, an individual must: pay a small annual fee; be employed by a registered company or business; and have completed an approved training course.[3]

In seeking to provide practical guidance to those working in the industry, RACIRB adopted an ozone code of practice specific to the refrigeration and air-conditioning industry. This had already been developed by the industry in New South Wales in the early 1990s, in order to create a common national standard, and, rather than attempting to 'reinvent the wheel', RACIRB chose to implement it essentially unchanged.

Enforcement

As with the administration of the registration and accreditation database, enforcement has been essentially devolved to the nominated issuing body, RACIRB. Consequently, DEP does not employ inspectors to address the use of ozone-depleting substances in the refrigeration and air-conditioning industry. Rather, the RACIRB Executive Director fulfils the role of 'quasi'-inspector; a role that is conducted in a number of ways.

First, the Executive Director conducts random (and unannounced) visits to business premises to inspect their operating practices and to provide advice and guidance where deemed appropriate. The intention is, at some point or another, to visit the premises of all registered businesses. The large number of registered business, their wide geographical dispersion and the limited inspectoral resources of RACIRB have, however, meant that it has taken about three years for this inspection programme to be completed.

Second, RACIRB operates a telephone 'hotline' for members of the community and/or industry to identify instances where the industry's code of practice has not been adhered to. For example, there have been situations where clients have reported hearing 'pronounced hissing sounds', which indicates that ozone-depleting substances may have been either deliberately or inadvertently 'vented' to the atmosphere. In such cases, the Executive Director will investigate the particular circumstances of the event, by visiting the work site and/or the business or individual involved.

Education and training

All workers in the industry must be accredited with RACIRB before they can be employed by a registered business, and in order to be accredited with RACIRB they need to complete education and training on the environmental aspects of handling ozone-depleting substances in the workplace. In Western Australia, all such education and training is conducted through the Technical and Further Education (TAFE) system. This is based in part on the national code of practice for the industry. For example, TAFE training provides participants with information on refrigerants and their properties, environmental issues,

---

3 In the first year of RACIRB's operation (as the official designated issuing body) in 1994, there were 662 registrations and 1,842 accreditations. By the end of 1998 these numbers had risen to 829 and 2,320 respectively. This demonstrates the large number of enterprises and individuals operating and working in the industry of Western Australia. This large number is a function of the dominance of small and medium-sized business in the sector, with even the largest companies rarely employing more than 50 people.

safety and evacuation issues, methods for replacing gases, and methods for the recovery of gases. The TAFE programme has a highly practical focus. Accreditation is open to anyone who chooses to complete the basic TAFE environmental programme (it should be noted, however, that no additional environmental training is providing in the full apprenticeship programme).

An evaluation

In determining the success or otherwise of ozone co-regulation in the refrigeration and air-conditioning industry it is not possible to utilise quantitative data. The very nature of ozone-depleting gases, which are odourless, colourless and non-reactive, means that it is very difficult to verify individual cases of venting, and credible quantitative data about the extent to which there is compliance with the regulatory requirements is simply not available. Nor is there any possibility of obtaining such data in the future. This is a common situation confronting regulatory agencies which, when regulating SMEs in particular, frequently have to make policy decisions on the basis of very incomplete information.[4]

In these circumstances one must rely on qualitative data and, in the particular case of ozone-depleting substances, this data is only obtainable from the groups that have direct experience of, and involvement in, the regulatory process. These, of course, include members of the industry itself, who are hardly likely to be impartial, and one might have particular concerns about the veracity of claims by industry representatives that they are regulating themselves effectively. However, as we will see, in the circumstances of ozone protection the accounts of industry members were very largely consistent with those of other informed players (including government regulators in particular) Moreover, industry members did not present a 'rose-tinted' picture of industry co-regulation. On the contrary, they were critical of some aspects of the current regime and anxious to improve it as part of a broader agenda of successfully 'professionalising' the industry. For these reasons, their accounts have greater credibility than might otherwise have been the case, and taken in conjunction with the high level of agreement between industry insiders and outsiders on potentially contentious questions, gave us confidence in the accuracy of the following evaluation.[5]

The main aim of the ozone-protection regime is to limit emissions of ozone-depleting gases through their correct use and disposal. Based on our interviews, the overwhelming

---

4   DEP will be in precisely this situation when it carries out its first formal evaluation of the self-regulatory programme for the control of ozone-depleting substances, shortly after the time of writing.

5   The fieldwork involved semi-structured interviews (26 in total) with a representative sample of stakeholders, including 18 enterprise managers, with the remainder including officials from government agencies, educational institutions and related professional associations. This included key officials of the major self-regulatory body, RACIRB, and the government regulator, DEP. Snowball sampling was used to identify individuals in the above categories most able to provide valuable insights. Interview material was supplemented by information from public documents, industry journals and reports.

majority of respondents from both within and outside the industry viewed the co-regulation regime, administered by RACIRB, as a success in achieving these objectives. For the most part, this view was based on their own observation of industry practice, on their broader interactions with industry members and stakeholders, and sometimes on a somewhat intuitive 'before and after' comparison. For example, an industry representative has claimed that:

> The role of the board [RACIRB] has been one of industry self-management to implement the requirements of the Policy. This has proved to be quite a successful model and there can be no doubt that the implementation of the current policy has led to a reduction in the amount of refrigeration being released to the atmosphere (Boyle 1999: 6).

And this view accorded with that of a government respondent who stated: '[Self-regulation] has been very successful. People now know what they are doing. I have no doubt as to the strength of this approach compared to those adopted by the Eastern States.'

Most respondents asserted that there was very little deliberate venting of ozone-depleting gases to the atmosphere, although industry members conceded that reports of customers hearing loud hissing sounds (implying venting) were not entirely unknown. Most industry respondents also acknowledged that they had witnessed examples of shoddy refrigeration and/or air-conditioning systems, and that such systems had almost certainly resulted in emissions of ozone-depleting gases. Nevertheless, while not claiming that current industry practices were ideal, the large majority of our respondents believed there had been a dramatic improvement since the introduction of co-regulation. A representative view of the policy's success was succinctly put by one industry respondent when he stated that:

> We know there are 'cowboys' out there. We all see their bad work. But the vast majority of businesses are all very switched on. Even though it could be used as a means to undercut the market, we now know the right way to do things, *and the reason why it's important*. Its no longer a major burden—the costs are simply passed onto the client [emphasis added].

Several respondents contrasted current behaviour with the *laissez-faire* practices of the past, when venting was common and there was little or no concern with the safe and environmentally responsible treatment of ozone-depleting gases. In this regard, it is salient to note just how far the industry has come in changing entrenched cultural attitudes towards the handling of refrigerant gases. Previously, these gases were treated by the industry as completely safe and environmentally benign. They are non-reactive, with excellent refrigerant properties, and relatively cheap and plentiful. It is not surprising, then, to hear of the cavalier way in which the gases were handled 'in the old days', prior to the advent of concerns about the declining stratospheric ozone layer. Indicative of this was a refrain from one industry respondent that 'the refrigerant R12 was commonly used to chill beer cans'.

We may conclude that co-regulation, in this instance, has been a qualified success—an assessment that does not sit well with the conventional theory on the application of self- and co-regulation as alternatives to command-and-control regulation in relation to SMEs. The possibility that self- or co-regulation might be a viable strategy for at least some SME sectors has been dismissed out of hand, for the following reasons:

- SMEs often lack the internal resources and expertise of their larger counterparts to successfully implement and manage internal environmental management systems which are often instrumental to the successful application of self- or co-regulation.

- There is no history of the refrigeration and air-conditioning industry having had a strong organisational structure, in the form of an industry association, necessary for effective management of a self- or co-regulatory regime.

- SMEs are far less likely to have a high degree of visibility within the community and therefore have less to lose (either politically and commercially) if found to be wanting in terms of meeting their self- or co-regulatory obligations—they cannot be shamed into compliance.

- The very nature of the potential environmental harm caused by refrigeration and air-conditioning companies and/or individuals not conforming to their co-regulatory obligations is largely undetectable, making it difficult to confirm compliance.

- There is no conventionally defined win–win solution involving both economic gain to industry participants and improved environmental outcomes.

Why, then, did the refrigeration and air-conditioning industry achieve better environmental outcomes under co-regulation than might have been anticipated? Why did the industry not succumb to the temptation associated with many such initiatives, of embracing self- or co-regulation to give the impression of putting one's environmental affairs in order, and to keep government regulation at bay, while lacking any serious commitment to environmental improvement? The main reasons would appear to be as follows.

First and foremost, there was a substantial coincidence between the public interest in environmental protection and the industry's own interests. Specifically, the overriding driver of the industry's support for co-regulation was the desire to grasp the opportunity for ozone protection to achieve wider industry objectives. In particular, many in the industry believed that it lacked central co-ordination and a high degree of professionalism, as well as quality management and operating standards. The need to introduce minimum standards necessary for registration and accreditation under an ozone-protection co-regulatory regime provided a solid base from which to build a more united, ambitious and influential industry structure, to improve the professionalism of the industry, and to 'raise the standards in the trade generally'. A common refrain was that:

> Before the existence of RACIRB, the industry lacked coordination and a united front. The need to address ozone depletion forced the industry to examine its professional standards and improve its levels of education and training. Without ozone, the industry would still be 'in the dark ages'.

In addition, it was suggested that the industry had a very poor business track record, with a very high rate of business failure, which was attributable, at least in part, to the fact that 'anybody was able to go out and set up a refrigeration/air-conditioning business'. Co-regulation forced the industry to establish an umbrella organisation to introduce industry-wide codes of practice, minimum levels of training and a system of accreditation. Thus the industry viewed its ozone obligations, and through it, the creation of

RACIRB, as a 'classic win–win, where the trade wins, the environment wins and the public wins'.

A second factor identified by many industry respondents (and also shared to a fair degree by government respondents) was that individual enterprises and employees were far more likely to comply with an 'enforcement' regime that was based on peer pressure and education and training than with one involving government inspectors and mandatory regulations. It was widely believed that an external regulatory agency would have little detailed understanding of the technicalities of refrigeration and air-conditioning. For example, many respondents doubted the potential efficacy of conventional regulation because of the difficulty in tracing gas leaks, the ubiquity of enterprises and the geographically dispersed and rapidly changing location of individual work sites.

Third, the enforcement regime established by RACIRB, while far from ideal, was at least credible (the absence of enforcement being one of the most fundamental failures of many self-regulatory initiatives). Although the Executive Director of RACIRB, in conducting his inspectoral responsibilities, had no legal powers of enforcement, this is substantially compensated for by using a system of 'bluff and persuasion' to bring about the desired changes in behaviour. For example, RACIRB's lack of formal power is not highlighted during site visits or widely communicated, and consequently many businesses and individuals are unaware of the Executive Director's legal impotence. In such cases, his mere presence at a work site serves to deter continued poor practices. Moreover, the fact that all enterprises must be registered with RACIRB assists its inspectoral role by 'giving it greater control over its membership base'. Members are made well aware of the power of RACIRB to withdraw registration and with it the right to continue operating.

Fourth, RACIRB is also able to draw on the authority of DEP in conducting its enforcement role. Unlike RACIRB, DEP does have the power to undertake various forms of enforcement action, including the power to prosecute breaches. This 'big stick' may be harnessed by, for example, inviting DEP officials to attend inspections with RACIRB, something that has occurred on a intermittent basis. DEP could also, in exceptional cases, contemplate prosecution, though it believes there would be serious evidential problems in achieving a conviction. Thus DEP has played important roles, lending authority to RACIRB initiatives and providing an important underpinning where such initiatives prove insufficient in isolation. For example, as a result of the close working relationship that exists between the Executive Director and DEP, the Executive Director's lack of formal legal authority has not been a major issue. Many of the respondents were very supportive of the constructive working relationship that had developed between DEP and RACIRB, and argued that this was an essential component in achieving improved environmental outcomes.

Fifth, the industry-initiated environmental training conducted by the TAFE also contributed to an improvement in standards and to providing new entrants to the industry with a heightened awareness of environmental responsibilities. Those closely involved with the administration of the course and the operation of RACIRB were strongly supportive of the training. They believed it facilitated a better understanding of the environmental consequences of releasing ozone-depleting substances, and provided the necessary practical guidance on how to minimise such emissions. However, even the strongest supporters of the training initiatives acknowledged that there were varied responses to the course, and that 'it is not possible to make the ideal work in the real world'. A common view was that those that had completed the TAFE module would be 'unlikely to be dumping CFC refrigerant, but that there may be shortcomings [in some of their other practices]'.

Sixth, peer pressure also contributed to both a heightened awareness of environmental issues and to improved environmental performance. There are many in the industry who do not welcome shoddy practices that give competitors an unfair competitive advantage and which may bring the entire industry into disrepute, and who are prepared to exert pressure on their peers in one of two ways. First, as noted above, RACIRB maintains a hotline that is available to industry, clients and the wider community to raise concerns about industry practices. In practice, most complaints are from fellow industry members. Once a complaint is received, RACIRB then acts to make an initial judgement as to its merits and, if warranted, will follow up with a site visit.[6] Second, there is a more subtle pressure for companies to ensure that their work is up to the standards of their peers. This was described as 'a form of professional pride, if you like'. Companies (and individuals working for companies) do not want to be exposed as having contravened the code of practice. They also acknowledge that they have a common interest in making self-regulation work (thus avoiding more onerous interventions). This, however, is likely to be more important for mainstream operators than those operating at the fringes of the industry, such as domestic split-system installers.[7]

Finally, the incentives to reclaim ozone-depleting substances created by the national reclamation programme has had a direct bearing on the success of the Western Australian initiative. Specifically, the national reclamation programme employs a price signal to promote appropriate practices in the industry. In short, enterprises and individuals receive a payment for each unit of ozone-depleting substance they return to a designated depot (in most cases, this will be a commercial supplier of replacement gases). The consensus view in the industry is that the amount of refund received roughly equates to the cost of the time and effort involved in correctly collecting the gas and returning it (as opposed to simply venting it). As such, it is still largely up to the goodwill of operators to abide by the industry codes of practice. So, while the national reclamation programme does not provide a positive incentive to 'do the right thing', it at least removes any disincentive to do so.

None of the above should be taken to imply that the co-regulatory regime was free from significant faults. On the contrary, both industry respondents and others identified a number of levels at which this approach had serious shortcomings, not least in terms of the range of enforcement options and the limited enforcement resources available, the lack of prosecutions, the absence of a paper trail and adequate monitoring, and the limitations of the current training regime. We deal with these limitations, and how they might be overcome, below.

6   Although it is claimed that this system has been reasonably effective in raising standards in the industry, it is not without problems. RACIRB notes, for example, that the majority of industry-generated calls do not arise out of a genuine concern for the environment; rather, they are simply a means of creating difficulties for rival operators.

7   It was suggested by some respondents that split-system installers are less likely to be influenced by peer pressure. Over the past ten years there has been an explosion of sales in domestic air-conditioners, the so-called 'split systems'. This has led to unqualified 'cut-price' installers operating: in many cases, sole trader businesses with unqualified staff (that is: although they are registered or accredited with RACIRB, they have not completed a full apprenticeship through the TAFE system). Many respondents suggested that they employed less than professional standards.

The need for further reform

Although qualitative data suggests that the RACIRB co-regulation programme has achieved considerable improvement in the environmental behaviour of members of the industry, there was no shortage of criticisms of individual elements of it, and little doubt that the programme as a whole could be improved. Below we discuss the main reforms that, if implemented, might substantially enhance its workings and achievements.

One of the most serious criticisms concerns the lack of adequate **enforcement resources**. Currently, there is only one RACIRB official (effectively part-time, given his other duties) responsible for monitoring, inspecting and enforcing the entire refrigeration and air-conditioning industry in Western Australia. With a large number of geographically dispersed refrigeration and air-conditioning operations, it has taken three years to visit each and every one of the registered businesses across the state. In the absence of greater resources, there is a serious risk that grossly inadequate government regulation is simply being replaced with grossly inadequate private regulation.

While it is arguable that co-regulation relies more on persuasion, education and training than on inspection and enforcement, the two approaches are complementary. Certainly a higher percentage will embrace voluntary action when the alternative is far less palatable (Ayres and Braithwaite 1992). Many respondents believe that the current low level of inspection is simply not sufficient to effectively 'enforce' the appropriate industry codes of practice, even with the judicious use of persuasion, bluff and (largely empty) threats of legal action. Provision for more inspectors would be a powerful way of 'getting the message across' to the industry and of supporting those managers trying to 'do the right thing' (by impressing on their employees the importance of good ozone practice). It would also facilitate enforcement action being taken against those companies and individuals that wilfully choose to flout industry guidelines and responsibilities. An increase in the annual registration or accreditation fee would be the most practicable means of providing for more resources, and, following some nudging from DEP, the industry seems poised to accept this option.

Even with greater inspectoral resources, the RACIRB inspector still lacks an adequate array of potential sanctions. This seriously undermines his or her ability to bring about change, particularly in those companies that may be less than enthusiastic about ozone protection. In short, the RACIRB inspector lacks regulatory 'teeth'. Ideally, regulatory sanctions should have the capacity to be imposed in a progressive fashion such that they can fit the size, nature and frequency of the breach in question. Braithwaite (1993) and others have argued that regulators must invoke enforcement strategies that successfully deter egregious offenders, while at the same time encouraging and helping the majority of employers to comply voluntarily, and that the best way of achieving this is by invoking an enforcement pyramid. At the base 'regulators assume and nurture virtue—corporate responsibility . . . When virtue fails, regulatory strategy shifts through escalating deterrent responses. When deterrence fails, strategy shifts again to an incapacitating response' (Braithwaite 1993: 88).

However, in the case of the co-regulatory regime administered by RACIRB, the capacity to escalate progressively up an enforcement pyramid is extremely limited. The RACIRB inspector has no formal sanctions at his disposal, and to the extent that persuasion coupled with bluff fails to achieve the desired result, the only option is to call in DEP to seek a prosecution. This is a very heavy-handed and costly exercise: in effect, it implies moving from the very base of the enforcement pyramid to its peak without any attempt to achieve more constructive solutions at the intermediate levels.

There are two potential opportunities to introduce a more progressive, and arguably more effective, movement up the enforcement pyramid. These are, in order of severity, on-the-spot fines and deregistration or removal of accreditation. In the case of the former, the RACIRB inspector has no legal capacity to impose such a penalty, although incidents that in his view merit such action could be referred to DEP for action. However, one of the attractions of on-the-spot fines (in most other contexts) is the ease with which they can be issued. The same is true for the various types of administrative notice that DEP has the authority to issue, but which, again, the RACIRB inspector does not. There is no ready resolution to this gap in the enforcement pyramid, save for the cumbersome approach of referring appropriate matters to DEP for action.

Moving up the enforcement pyramid: if enterprises or individuals were demonstrably and repeatedly recalcitrant, despite the imposition of lower-order measures, then RACIRB inspectors could be given the authority to refer individual cases to RACIRB for deregistration or removal of accreditation.[8] This would be a very serious step, as it would effectively prevent the company or individual from operating in the industry (they cannot get access to replacement refrigerant gases or products containing those gases without official registration or accreditation). It would, however, be a more useful sanction than prosecution because, again, such action could be taken by RACIRB itself, and it would not require the imposition of legal proceedings by DEP. It could therefore be left to the discretion of RACIRB to make a judgement on the merits of each case, thus freeing it of the need to gather absolute legal proof. The option of prosecution would still be retained where it was considered justified, and where there were reasonable prospects of a conviction being obtained.

A further flaw in the operation of the existing enforcement regime is that it is questionable whether the threat of prosecution it a credible one. Even though the Executive Director of RACIRB has no legal powers of enforcement, he and DEP co-operate closely on enforcement issues and it would be possible, albeit difficult in evidential terms, for DEP to launch such prosecutions in appropriate circumstances. No such prosecution has ever been brought, and there is a risk that in the long term the entire enforcement regime may come to be perceived as toothless, and thus become increasingly ineffective. For this reason, several respondents suggested the Department should 'try harder' to achieve a successful prosecution, because this would send a powerful signal to the poor operators in the industry, and because it would support the overwhelming majority of companies and individuals that are 'doing the right thing'.

An additional concern relates to monitoring. The absence of quantitative or even indicative measurements by which it is possible to make a considered judgement about

8   The use of industry sanctions, such as deregistration and removal of accreditation, raises the question of 'anti-competitive behaviour'. According to the (former) Department of Industry, Science and Tourism, 'where industry has the commitment to collectively sanction breaches of code there is the possibility that such action may amount to anti-competitive behaviour. In most cases such action may not amount to anti-competitive behaviour or the benefit to the public may outweigh such behaviour. However, to avoid any threat of legal action for breach of the competition provisions of the Trade Practices Act 1974 (Cth), a procedure exists whereby industry can have the arrangement authorised.'

the programme's success and/or progress is not merely a problem for researchers but also represents a potentially serious flaw in the operation of the self-regulation programme itself. If RACIRB (and to some extent DEP) is not in a position to determine the shortcomings and/or positive features of the co-regulatory programme, then it cannot make considered judgements on the allocation of resources in the future, or identify opportunities for policy fine-tuning.

How might better monitoring be achieved? The best option would appear to be the creation of a viable paper trail. Under existing arrangements, no record of individual jobs is kept. Thus a valuable source of information on the amount and types of ozone-depleting gases used or recovered is lost. Such information is not only vital for determining the merits of individual cases of potential breaches of the industry code of practice but, importantly, it could be used to build up an industry profile of ozone-protection management practice norms across particular applications. Although not foolproof, the mere presence of a paper trail would act as an incentive for both companies and individuals to be more vigilant. In order to be successful, however, a paper trail would need to be easy to implement, and not create an unrealistic administrative burden on operators. One suggestion from several respondents was for a simple job description card to be completed for each job. This could be linked to records of payment for each job, and made a compulsory component of annual registration.

It might also be possible to utilise the existing reclamation arrangements to track the performance of individual companies. Under existing arrangements, companies can only purchase replacement gases from approved outlets, and they must return ozone-depleting substances to the same outlets. There is thus an opportunity to keep records of purchases and returns by individual companies and, importantly, to match these records against industry norms. Thus those companies found to display abnormal purchasing or return patterns could be further investigated by a RACIRB inspector.

There were also concerns about training. Although there was widespread support for the current initiative in training industry employees through the TAFE system, in practice this is far from ideal. A substantial minority of industry respondents, for example, bemoaned the general poor quality of qualified tradesmen and -women exiting the TAFE system. The perception was that the general level of environmental awareness that was imparted to TAFE students, while adequate, could be substantially improved. At the very least, it was suggested that DEP could expand its educational contribution by participating in the TAFE training programmes of industry employers and employees such that 'environmental awareness' is generated, and the programme itself could be encouraged to include a more substantial environmental component.

Finally, the concept of co-regulation implies that the government ensures that the industry self-management, and the broader parameters of the regulatory regime as a whole, satisfy a number of criteria. For example, it is clear from an increasing number of empirical studies of self- and co-regulation, and from voluntary approaches more generally, that the following are of central importance to regulatory design: that the policy goals of the regulatory regime are developed by means of a transparent process; that clear, measurable targets are stipulated; that incentives are provided for participation; that there is provision for monitoring of results (by third parties if possible); and that there are adequate sanctions for non-compliance.[9] In the present context, it should be

---

9   See, in relation to voluntary approaches, Moffet and Bregha 1999.

noted that it has been difficult to set clear and measurable targets; that, while disincentives have been removed, incentives for individual operators (as distinct from those for the industry association) are lacking; and that, to date, there has been no monitoring of results (a matter that, as we have indicated, presents an enormous challenge); and that the issue of effective sanctions remains problematic for reasons identified above.

Policy implications

Beyond the immediate concerns of this particular industry sector, what wider policy implications does this case study have for achieving improved environmental outcomes in the small-business sector generally? The broader lesson is that seemingly unorthodox applications of regulatory and quasi-regulatory policy instruments may prove to be viable options for SMEs. In particular, and contrary to the conventional wisdom, there may yet be a useful role for co-regulation and industry self-management to play in regulating SMEs, in at least some contexts. But what are those contexts likely to be, and, within those contexts, how must co-regulation be designed in order to achieve success? Indeed, what does success actually mean, in the context of regulating SMEs?

In what circumstances are self- or co-regulation credible policy options?

As noted earlier, the academic literature cautions against the application of self-regulation (or co-regulation) where there is, *inter alia*: no track record of co-operation with government; no experience with self-imposed environmental constraints; little exposure to consumers' environmental values; low corporate 'visibility'; numerous corporate players; high exit costs; and no recognised industry association. These criteria are arguably characteristic, to a greater or lesser extent, of many SME sectors, and certainly apply in the case of the refrigeration and air-conditioning sector. Yet this case study demonstrates that, despite such requirements, environmental co-regulation may still be a viable policy option for a SME sector *provided there is a powerful self-interest at stake*, such that the industry or industry association's goals are reasonably consistent with the broader public interest in environmental protection (so called win–win outcomes).

To an extent, the environmental policy literature already recognises the capacity to exploit opportunities to achieve win–win outcomes. The cleaner-production literature, in particular, suggests that enterprises will be willing to implement many environmental improvements voluntarily, provided they are aware of the economic advantages that may accrue (see e.g. Gunningham and Sinclair 1997). The extent to which such economic and environmental win–win outcomes exist is still a matter of considerable debate, although it is generally agreed that much more low-hanging fruit remains to be picked by SMEs than by large enterprises. However, there are also circumstances in which such win–wins are demonstrably not available and where such voluntary initiatives are unattractive to industry. Ozone protection is one such area because, at best, industry operators can break even by reclaiming and returning gases to the authorities for appropriate disposal.

What this case study reveals is that the conventional definition of win–win, in terms of individual economic benefits to enterprises from environmentally responsible behaviour, is far too narrow. As we have argued, in the case of the refrigeration and air-conditioning industry, raising the standards of training, business practices and overall professionalism were the stated goals, not improvements in productivity and, with it, conventional economic gains. That is, while co-regulation is likely to be viable only where the relevant industry and/or its association have a self-interest in its establishment, this self-interest need not be a purely economic one. It may include, for example, better community relations, higher standards of industry professionalism, an enhanced corporate image, and forestalling the need for government intervention. All of these broader industry interests may be exploited by policy-makers seeking to design viable co-regulatory programmes.

However, the presence of such a self-interest must be widely recognised and acknowledged. There is little advantage in the existence of a genuine self-interest if most in the industry are ignorant of its presence, or are unconvinced that it is sufficient to justify changing their behaviour. In these circumstances, the risk of free-riders is substantial. Moreover, there must also be *a demonstrable link between an industry self-interest and the introduction of co-regulation*, before such action is likely. In the case of the refrigeration and air-conditioning sector, the industry came to recognise that the *very act of co-regulation* necessarily required a much higher degree of industry organisation, training and co-operation, and this could lead to the much sought-after improvements in industry practices and professionalism. In this particular instance, it is arguable that few other policy instruments could have achieved the same outcome, and certainly not with the same degree of success.

However, it was only the threat of interventionist ozone-control regulation that raised this issue to sufficient prominence to make this link clear in the first instance. The broader issue highlighted here is that, in the absence of a credible threat of direct government intervention, and associated loss of autonomy and additional costs, the possible attractions of co-regulation would not have been sufficient to generate industry action. Above all else, it was the threat that, if the industry did not act, government would impose some far less attractive form of regulation on it, that galvanised the industry into action. This point should not be forgotten, for as the broader self- and co-regulation literature makes clear, credible industry action in the absence of such a threat is very uncommon.

While the above circumstances may predispose industry to attempt to genuinely regulate itself, they do not guarantee that such efforts will be successful. There is still a need to overcome the structural and institutional deficiencies of such a small-business sector in seeking to successfully *implement* self-regulation. It is extremely doubtful that such an SME sector would be in position to address these issues without some form of external support.

Co-regulation and self-management

As noted in earlier, the refrigeration and air-conditioning industry's ozone strategy in Western Australia is more accurately described as co-regulation than self-regulation. Even though the industry, through RACIRB, has administrative control of the programme, and has significant policy input, it is DEP that retains the legal power of enforcement and, indeed, legislated (albeit with significant industry input) the policy goals that the RACIRB initiative seeks to achieve.[10] That is, the strategy involved government target-setting in conjunction with industry association self-management. And, even in respect of the latter, RACIRB's enforcement action and credibility required an underpinning of government support, without which the industry association could not have achieved its environmental goals. As noted above, RACIRB regularly draws on the authority of DEP both explicitly and implicitly as it goes about its daily business of implementing self-regulation in the industry.

The precise government role in such schemes will vary from case to case. For example, government may engage directly in the co-regulatory process by jointly negotiating targets and strategies with industry, and by providing external verification and/or ratification. Government may also be involved to varying degrees in education and training, in providing financial subsidies and other incentives, in providing preferential procurement or contracting policies, and, crucially, in supplying an enforcement 'safety net' if and when industry self-regulatory provisions prove insufficient. In this last case, government intervention may be at the invitation of the designated industry association, or based on a unilateral decision. Finally, government must delegate regulatory responsibility to a designated industry association, and, where such an organisation does not yet exist, take active steps to create one. In all cases of government support, a productive working relationship between the industry association and the government is essential.

The slightly different terminology of 'regulatory self-management' is used to describe a policy process whereby 'government sets regulatory policy and rules by legislation: the self-management organisation or firm is responsible for operational delivery of the program or implementation rules' (Priest 1989–99). Further, 'management powers [are] delegated by government through contractual agreement', 'government will monitor the performance and conduct of [the] industry organization' and adjudication is 'initially through discipline/dispute settlement mechanisms in [the industry] organization' (Priest 1989–99). Again, although arguably somewhat more formal and sophisticated, this is largely consistent with the approach adopted in Western Australia to control the use of ozone-depleting substances by the refrigeration and air-conditioning sector.

A final policy issue is whether co-regulation is the exclusive domain of SMEs and government. In other words, might there be other organisations or institutions that could fulfil the role of government in overseeing, supporting and, where necessary, directly intervening in the self-regulation of small business? It is possible in fact that there are a range of third parties that could act as surrogated co-regulators. For this to be effective, however, they would need to have the capacity to wield, if necessary, enforcement sanctions at least as effective as those at the disposal of government. No third-party surrogates were available in the case of ozone protection: neither environmental groups nor large suppliers could credibly have performed this role.

In other circumstances, however, there may be considerable scope for leverage via much larger and more powerful business entities on whom SMEs are dependent for their livelihood. This would include, for example, smaller parts suppliers to a major automotive manufacturer. In such instances, the dominant commercial entity successfully enters

---

10   One benefit may be a greater commitment to abiding by regulations. Experience in the Netherlands, for example, indicates that the very act of negotiating co-regulatory agreements provides industry with a greater insight into, and commitment to, environmental management.

a co-regulatory arrangement with its much smaller suppliers. There have already been tentative steps in this direction, for example, in the case of chemical industry's Responsible Care programme. The possibility of such an arrangement, however, does not address the question of *why* such an arrangement may come about. It may be that an initial impetus from government is necessary, as described above, even if there is no ongoing government engagement.

'Success' as a relative concept

In seeking to design means of regulating SMEs, policy-makers are confronted with grossly imperfect options. There are usually vast numbers of such enterprises to be regulated yet only extremely modest resources available to do so. In this context, co-regulation should not be judged against some abstract and theoretically ideal strategy but rather in terms of how it 'stacks up' against other *practicable* policy options. When viewed from this perspective, the alternatives to co-regulation are often neither very many nor very attractive.

One policy option would be to introduce a conventional command-and-control regime that mandated particular controls on the handling of ozone-depleting substances in the refrigeration and air-conditioning industry, and provide sanctions for any deviation from these standards. There are a number of obvious deficiencies in doing so. First, it would require substantial inspectoral resources to effectively enforce the mandated controls across a large number of companies operating in a variety of changing work sites. Second, government inspectors are generalists rather than specialists and lack the expertise to appreciate the complexities of at least some industry-specific environmental problems. Third, and arguably most importantly, effective controls on the use of ozone-depleting substances in the industry required, fundamentally, a change in individual attitudes. It is doubtful that an externally imposed enforcement regime would be successful in this regard, particularly when compared to a self-regulatory regime that had attracted widespread support from within the industry.

Another policy option might have been to introduce a control strategy based on the application of economic instruments. There are two possible contenders in this regard. First are tradable permits. In this instance, each company would either be provided with, or have to bid for, a certain number of 'rights' to purchase and use ozone-depleting substances (or their replacements) which they could then trade with other companies. Briefly, this would be wholly impractical to implement in light of the large number of companies operating in the industry, and the need to monitor every trade that took place. And there is no guarantee that controls on the *supply* or replacement of ozone-depleting refrigerant gases would necessary prevent the inappropriate handling of ozone-depleting substances.[11]

The second option is the use of price signals. This could be a negative price incentive, in the form of a tax, or a positive one, in the form of a subsidy. In the case of the former, it is not obvious at which point a tax could be realistically imposed to bring about the desired changes in behaviour. For example, a tax on the sale of ozone-depleting gases (ignoring for the moment the fact that the Commonwealth was already introducing

---

11   In any case, the Commonwealth had already introduced a national tradable cap scheme at the level of company import and manufacture.

supply controls) would do little to effect the use of *existing* gases in the myriad refrigeration and air-conditioning systems already in operation, apart from greatly increasing their value, and potentially preventing their safe return by creating a black market. In the case of the latter, one option would be to offer the industry a payment for the safe return of ozone-depleting substances to designated sites. This has the potential to be an effective strategy, but its attraction would have been diminished by the following two factors: it would have the potential to become a significant strain on state resources (thus contravening the 'polluter-pays principle'), and the Commonwealth was already instituting a national reclamation programme, thus rendering state-based efforts duplicative and potentially obsolete.

A more courageous policy strategy might have been to avoid a controls-based approach altogether, whether mandated through legislation or effected by economic instruments, and instead rely on a comprehensive programme of education and training, on the assumption that the key to success is changing the attitudes and behaviour of companies and individuals operating and working in the industry. However, to the extent that this is correct a key question that arises is who is best placed to deliver the message to industry? History suggests that an education and training programme emanating from the industry itself, or at least under the guise of an industry-based co-regulatory programme, would have more credibility and be more readily received from within the industry than a purely government-based programme (see Chapter 2).

An important insight that arises from this last point is that not only were policy-makers choosing between less than perfect options, but those options, in many instances, are not mutually exclusive. For example, education and training is a natural bedfellow of virtually all other policy strategies, and arguably has a particular affinity for self-regulation. Economic instruments, too, might be effective partners to self-regulation. For example, offering payments for the safe return of ozone-depleting substances is also likely to support self-regulatory efforts. On the other hand, a fully fledged command-and-control programme would self-evidently be in conflict with a programme of self-regulation. Thus an important consideration for policy-makers was not just how a policy strategy for the refrigeration and air-conditioning industry might operate in isolation, but *how it would interact with other policy strategies* (either existing or contemplated).

The preceding discussion does not establish that the co-regulation programme implemented by DEP was optimal. But it does demonstrate that the programme incorporated a sensible mixture of policy components (it was underpinned by the threat of government enforcement, it incorporated, through central-government strategies, the removal of economic disincentives to complying with regulatory requirements, and it was integrated with a training and education strategy). And it did achieve, according to qualitative evidence, some considerable improvement in industry environmental performance. It was, at the very least, a viable policy approach, and it may well have been (though we cannot be sure since there is no immediate point of comparison) the best available option.

Conclusion

Notwithstanding the serious doubts expressed by previous writers about the effectiveness of co-regulation in relation to SMEs, our qualitative data suggests that it proved to be a viable policy option in the case of ozone protection in the air-conditioning and refrigeration industries. Both industry awareness and environmental outcomes improved substantially under this approach, and, while it certainly did not serve to eradicate shoddy practices and all emissions of ozone-depleting gases, deliberate venting seems to have been largely eradicated and overall environmental practices greatly improved.

Three broader lessons can be derived from this case study. First, co-regulation is more likely to be successful if the industry chooses to embrace it as a means of achieving broader, but complementary, objectives (giving rise to a sufficient coincidence between the private interests of the industry and the public interest in environment protection). An important role for public policy-makers is to harness that enlightened self-interest and in so doing both minimise the use of scarce public regulatory resources, and maximise the capacity of industry to be part of the solution, not just part of the problem.

Second, such a regime requires a judicious combination of policy instruments to ensure its success. These include the removal of economic disincentives to the reclamation and return of ozone-depleting substances, an industry training and education programme to teach industry operatives why ozone protection is important and how to achieve it, a system of certification and accreditation without which it is not possible to do business, effective monitoring, and credible enforcement (including an adequate range of enforcement tools). Crucial in reinforcing all of the above will be peer-group pressure and an underpinning of government regulation where all else fails.

Finally, in evaluating the success of any policy initiative to curb the environmental excesses of SMEs, it is crucial to remember that the available choices are extremely limited, and that one is usually comparing grossly imperfect policy options. Hence one can conclude that, contrary to the conventional wisdom, there are a range of circumstances where the seemingly unorthodox use of industry co-regulation and industry self-management may be a viable, and sometimes even the only viable, policy option.

Postscript: future directions

As noted throughout this case study, one of the striking features of the self-regulatory programme adopted by the refrigeration and air-conditioning industry in Western Australia is the inextricable link that has been created between the designated industry association, RACIRB, its desire to improve professional standards across the industry, and the achievement of predetermined environmental imperatives: namely, ozone protection. RACIRB owes its entire existence to the presence of this overriding environmental objective. This has serious implications for the longevity of RACIRB.

As the productive life of existing ozone-depleting substances and equipment comes to an end, and is either retrofitted and upgraded, or replaced by new non-ozone-depleting substances and equipment (as is rapidly occurring), the future role of RACIRB inevitably comes into question. The industry can foresee a day, in the not-too-distant future, when controls over the use of ozone-depleting substances will no longer be a significant policy issue. At this point, the *existing* legislative and policy rationale for RACIRB's existence will collapse. Some in the industry view this with concern, and fear it will trigger a return to a much more fractured and unprofessional sector lacking policy co-ordination and a strategic direction.

In light of this arguably bleak prospect (at least from the perspective of RACIRB), the industry, and RACIRB in particular, is looking for a potential policy 'saviour' in order to maintain the wider, non-ozone-specific benefits that have accrued to the industry. As such, it has latched onto a prospective candidate in the form of climate change. As with depletion of the stratospheric ozone layer, climate change is a global phenomenon associated with the accumulation of anthropogenic gases. In this case, instead of ozone-depleting substances it is greenhouse gas emissions. The most common of these are carbon dioxide (associated with the burning of fossil fuels) and, to a lesser extent, methane (associated with agricultural production and land-use changes).

As it turns out, existing ozone-depleting gases, and some of their non-ozone-depleting replacements, are potent greenhouse gases. It is not difficult to imagine, therefore, how the existing self-regulatory role of RACIRB might be transposed from ozone-depleting substances to the effective management of greenhouse gas emissions. In conclusion, it seems that not only was the environment instrumental in the creation of RACIRB and the wider professional gains of the refrigeration and air-conditioning industry in Western Australia, it may hold the key to its survival into the future as well.

5

CLEANER PRODUCTION AND THE VICTORIAN VEGETABLE GROWERS

SMEs are a rapidly growing and increasingly important component of industry whose aggregate environmental impact, in some respects at least, may substantially exceed that of large business. The strategies for curbing the environmental excesses of SMEs are as yet under-developed. No systematic policy solutions have so far been identified by policy-makers or regulatory strategists.

However, what is clear is that the conventional regulatory response that has been used to deal with large point-source polluters—command-and-control regulation—is singularly inappropriate for dealing with the vast numbers of SMEs that populate some industry sectors, which are manifestly beyond the reach of the small and diminishing numbers of state regulators.

This case study of the vegetable-growing industry in Victoria examines the limitations of conventional strategies and how these might be overcome through the use of cleaner-production partnerships and supply chain pressure.[1] It also argues the need to adopt a complementary mix of policy instruments in order to regulate effectively the diversity of circumstances and motivation of different regulated enterprises.

---

1   The fieldwork involved semi-structured interviews (15 in total) with a representative sample of stakeholders, including nine enterprise managers, with the remainder including officials from industry bodies, government agencies, research organisations and related professional associations. This included key officials from the major industry body, VVGA, and the government regulator, VEPA. However, given the small size of the industry body (two full-time staff), and the fact that one VEPA official serviced the region, this did not translate to a large number of interviews. Snowball sampling was used to identify individuals in the above categories most able to provide valuable insights. Interview material was supplemented by information from public documents, industry journals and reports.

The context

A common feature of major Australian cities is the presence of a substantial number of commercial vegetable growers (commonly referred to as 'market gardeners') at their suburban fringes. This is where the urban population meets rural industry. Vegetable growers have chosen such locations for obvious reasons. With product freshness a key consumer demand, they need ready access to their major markets, which include the major supermarket chains, fruit and vegetable markets and food outlets. With environmentally sensitive urban populations in close proximity, however, the activities of the vegetable growers have come under increasing scrutiny and their environmental management practices have often been found wanting.

Three major environmental problems confront the vegetable-growing industry across Australia, and have become concerns for proximate urban populations. These are: farm run-off of chemical pesticides and fertilisers into surrounding waterways (this is a likely source of nutrient contamination flowing from creeks into coastal bays); chemical spray-drift from aerial crop-dusting (which may have health implications as well as environmental impacts); and unpleasant odours arising from the use of organic manure. Other potential environmental issues include the storage and disposal of farm chemicals, water usage and energy usage.

The tension between suburbia and the vegetable growers caused by these environmental problems has not gone unnoticed by environmental regulatory agencies. In recent years such agencies have increasingly turned their attention beyond more traditional and larger point-sources of pollution (such as large factory sites, which are now substantially controlled through licences and other mechanisms). In particular, they have begun to focus on more diffuse and smaller non-point sources of pollution such as farms, which have not traditionally been subject to regulation, but whose environmental impacts are considerable.

This case study examines the Western Port region of Victoria where, in the face of growing community concern about pollution run-off from farms, the vegetable growers and the Victorian Environmental Protection Authority (VEPA) have formed a partnership to introduce an environmental management system throughout the industry. We begin by describing briefly the nature of the vegetable-growing industry and the major elements of its proposed cleaner-production partnership with VEPA. We then consider the policy insights and implications of this case study.

The industry profile

Vegetable growers supply the major supermarket chains, fruit and vegetable markets and restaurants and other food outlets with fresh vegetables for sale directly to the consumer, and food manufacturers with fresh vegetables for further processing: for example, frozen or canned vegetable products. A key feature of the industry, which distinguishes it from many other agricultural sectors, is the variety of produce it generates. While many other farmers tend to specialise in less than a handful of different commodities—for example, wheat, cotton or canola—vegetable growers produce a plethora of different products to meet diffuse market demands and to accommodate to seasonal variations.[2]

In Victoria, the vegetable growers number approximately 600. Of these, around 500 are located in the wider Port Philip Bay–Western Port region just to the south and south-east of Melbourne. Vegetable-growing enterprises vary greatly in size and in management sophistication. The vast majority may be classified as small to medium-sized businesses (fewer than 50 employees and 100 employees respectively). Most numerous are the very small, family-run operations that rely on only a handful of permanent staff backed up with seasonal contract labour. In such cases, few owner-operators have specialised management, marketing or financial skills; nor do they have the resources to employ this expertise, and are therefore forced to cope as best they can, often without a systematic approach to farm management. In contrast, the minority of larger operations are in a position to appoint specialised management staff, and consequently have a far greater tendency to employ relatively sophisticated management processes.

This dichotomy of size and management style is reflected in the divergent grower membership of the relevant industry body, the Victorian Vegetables Growers' Association (VVGA). Of the 600-odd growers, only about 150 are members of the VVGA and, of these, the larger, more sophisticated growers are predominant. As one industry respondent bluntly stated: '[Members of the VVGA] are the leaders of the industry . . . the "complainers" of the industry do not get involved in the VVGA.'

Like many other industries in Australia, particularly those with large numbers of SMEs, the vegetable growers have undergone a period of rationalisation in recent years as the larger, more sophisticated operations gain market share through greater economies of scale. Tight profit margins have made it difficult for smaller growers to compete. A powerful illustration of this transformation of the industry was provided by one respondent, who stated that: 'six to eight years ago, there was a four- to five-year waiting period to get a grower or seller stand at the Footscray markets [a common outlet for smaller growers]. Now, over a hundred stands are available with no one to fill them.'

However, unlike many other agricultural sectors in Australia, the vegetable-growing industry lacks a co-operative structure, such as exists in the rice, dairy or wheat industries, for example. In part, this is due to the fact that the industry does not produce a single commodity, which would lend itself to a co-operative sales and marketing approach for either domestic or export markets, and in part to the lack of a culture of co-operation in the industry. As one industry respondent noted: 'Networking is not a strong feature of the industry . . . with many growers actively undercutting each other's prices. Previous attempts to organise industry networks, along the lines suggested by government policy, have failed.'

---

2   Even those farmers focusing on what would commonly be considered a vegetable product—for example, potatoes—are not considered part of the industry on the very basis of this specialisation, and indeed have a completely independent industry association.

The cleaner-production partnership

Like an increasing number of agricultural sectors across Australia, including the beef, cotton and horticulture industries, the vegetable growers of Victoria have determined to develop a strategy for improved environmental performance. What distinguishes the vegetable growers' approach from many of the others is that they are attempting to achieve this in 'partnership' with VEPA. In this section, we first describe the key elements of this partnership, and, second, what forces led to its inception.

The VVGA has received joint funding from VEPA and the Horticulture Research Development Corporation (which draws on an industry-derived levy) to develop and implement an Environmental Improvement Plan,[3] with an emphasis on cleaner production. To this end, an agreement has been signed between VEPA and the VVGA which outlines the nature and scope of the plan, and a steering committee of industry and government representatives has been established to oversee its implementation. The purported aims of the plan are: to better understand the real impacts of market gardens on the environment; to increase growers' awareness of their environmental responsibilities; to provide a venue for growers to demonstrate good environmental performance; to reduce compliance costs; and to satisfy regulators and the community that the vegetable industry is environmentally aware and responsible. It is also anticipated that, by taking the initiative, the growers will avert the threat of tougher new regulations being introduced.

It is intended that the environmental improvement plan begin with a pilot phase, involving a limited number of vegetable-grower participants. There are four key elements to this phase: a pollution audit; an awareness programme and management audit; environmental management guidelines; and a training programme.

The first element of the pilot phase is to conduct an audit of participating farms to determine the level of pollutants in farm run-off. This audit will take the form of water samples, both before and after storm events. The purported aim is to determine the veracity of claims that the industry is a major source of pollution, in the form of pesticides and fertilisers, in Western Point Bay tributaries such as Watsons Creek. Other environmental issues including spray-drift and odour are not being subject to independent testing at this stage.

The second element of the pilot phase is to conduct a more general environmental management audit of participating vegetable growers and to raise the level of environmental awareness across the industry. Both these components are intended to address the full gamut of potential environmental issues, not just polluting farm run-off. They will include, for example, issues such as the storage and use of chemicals, energy use, remnant vegetation and biological diversity, spray-drift and odour. More broadly, the purpose of the management audit is to provide the necessary practical background from which it will be possible to design a management system that is tailored to the needs and circumstances of the vegetable growers. It is not intended that the management audits be used as compliance audits, and, as such, the results of each individual farm audit will remain confidential.

---

3 An Environmental Improvement Plan is a set of targets for reducing the environmental impact of a specific industry or company, and negotiated in consultation with external stakeholders. See VEPA 1993b and Chapter 8 below.

The third element of the pilot phase is to synthesise the findings from the previous elements, together with the examples set by other domestic and international agricultural sectors, into a set of environmental management guidelines tailored specifically to the vegetable growers. In this respect, particular attention will be focused, domestically, on the experiences of the cotton, beef and horticulture sectors, and, internationally, on the Linking Environment and Farming (LEAF) programme in the United Kingdom.[4]

In effect, the guidelines will act as a form of self-audit. At this stage, it is not intended that they be subject to external verification or accreditation. Nor is it anticipated that they will conform to established environmental management systems standards, such as ISO 14001, at least initially. Instead, the focus is on producing highly practical and easily understood guidelines that have the maximum chance of uptake by the majority of growers. As such, it is more accurate to describe the expectation as being one of 'good practice', rather than necessarily 'best practice'. As the VVGA emphasises: 'The industry needs to develop a simple mechanism by which the growers can demonstrate good environmental performance.'

If this exercise is successful in achieving a transition to good environmental practice among pilot enterprises, the next phase will be to market the programme across the vegetable-grower industry as a whole. At this point, it is anticipated that this will be achieved via an education campaign in combination with a series of workshops or training sessions.

The final element of the pilot phase is to develop and conduct a series of training sessions for vegetable growers to 'demonstrate the application of the guidelines'. Presumably, the intention is that individual growers will become sufficiently interested and sufficiently familiar with the environmental management guidelines to implement them on-site. To this end, suitable materials will also provided to participants. Given the inevitable time constraints of vegetable growers, the training sessions are likely to be kept relatively short.

Policy analysis

This case study presents an interesting example of an agricultural industry that is at the very early stages of taking stock of its environmental image and performance, and of developing a strategy for improving them both. It also provides a valuable insight into the changing nature of the relationship between a government regulator and an industry sector as the former attempts to, in effect, transform itself from an old-style command-and-control regulator into a facilitator of industry environmental self-regulation (in addition to providing a regulatory safety net). With this in mind, the following policy analysis of this case study is addressed under two headings: **key drivers** (of both the industry's and VEPA's engagement to date) and **future challenges** (which may have to be overcome to ensure long-term success).

4 www.leafuk.org/LEAF

Key drivers

The first question of interest to policy-makers is why the vegetable growers of Victoria have decided to systematically address its environmental performance. A related second question is why, in particular, they chose to do this through the vehicle of a 'partnership' with VEPA. In terms of the first question, there are four key drivers of industry engagement: community pressure; international trends; domestic trends; and future regulations.

In Victoria, questions about the environmental performance of the vegetable growers have focused on Watsons Creek, a tributary to the greater Western Port Bay. The region around Watsons Creek, as noted earlier, has seen a large influx of suburban residents in recent years, and is in fact one of the fastest-growing regions in Victoria. A voluntary conservation group, the Western Port Action Plus Group, has been established in the region and reflects the changing nature of community priorities. The Western Port Action Plus group has a broad stakeholder membership, including local residents, Landcare[5] workers, VEPA and industry representatives.

This has been the primary vehicle through which concerns have been raised about polluting run-off from vegetable growers. It was at this forum, in particular, that the industry was directly confronted with community opposition to its environmental impact, particularly run-off, and was subject, as one industry respondent put it, 'to a lot of finger-pointing'. Subsequently, the industry realised that, irrespective of the accuracy of such perceptions, they were a genuine threat to its long-term sustainability. According to one industry representative, the views of this group are symptomatic of wider trends in community attitudes:

> In the longer term, the industry can only operate if it has the support of the community . . . [the Western Port Action Plus group] has provided a catalyst for industry action. [We have] got to be *seen* to do something as well as doing it [emphasis original].

It was in this context—indeed, at a meeting of the Western Port Action Plus Group— that the representatives of the vegetable growers approached VEPA to explore possible mechanisms for addressing their environmental performance. In response, VEPA offered to provide funding for an industry-based initiative through its cleaner-production partnership programme.

Community pressure, then, can be identified as a key driver of industry participation. However, it was certainly not the only such driver. The vegetable growers had already been sensitised to questions about environmental performance—in particular, the use of environmental management systems and industry self-regulation—through both international and domestic trends. On the international front, a delegation of industry representatives had recently visited North America and the UK. There they were exposed first-hand to the dramatic changes in purchasing policy being introduced by major supermarket chains, a vitally important commercial market for vegetable growers. Industry representatives cited the example of Sainsbury's, which gives a clear preference to growers with strong environmental credentials who, crucially, can demonstrate this by conforming to predetermined environmental management standards. Canadian vege-

---

5   A national voluntary environmental improvement programme in Australia that links farmers and local communities.

table growers are also introducing environmental management systems, for broadly similar reasons.[6] This international visit crystallised the potential for similar domestic commercial environmental pressures to impact directly on the Australian industry in the future. As one grower pointed out: 'They [vegetable growers] are also concerned that retail chains may in the future develop unrealistic expectations relating to environmental performance and expect growers to comply with harsh requirements.'

On the domestic front, industry representatives had become aware of moves within other agricultural sectors to introduce environmental management codes of practice. A major influence, in this regard, was the Best Management Practice initiative of the Australian cotton industry. Another source of influence were the activities of the Western Australian government's agricultural department, Ag West. This organisation had come up with SQF 2000, the quality assurance standard adopted by vegetable growers across Australia, and had recently begun to incorporate environmental management practices into its standards. Such moves had not gone unnoticed by the VVGA:

> Other agricultural industries are already taking steps to demonstrate their commitment to the protection of the environment. The cotton and the beef farming industries have already produced guidelines to ensure their constituents are aware of the practices that will minimise negative impacts on the environment. The horticultural industry in Queensland has produced a code of practice for sustainable fruit and vegetable production. The vegetable industry must take a proactive approach to ensure that the commitment to the protection of the environment is demonstrated.

Interestingly, one potential factor that apparently has had little bearing on industry motives is the *existing* application of environmental regulations, or, more particularly, a desire to ease some of the current regulatory burden. This is understandable. The vegetable growers do not belong to a licensed industry, and most have had little or no contact with VEPA inspectors. In fact, the relevant regional VEPA office has only one inspector allocated (part-time, with responsibilities for several other industries) to the vegetable growers. Yet that official is responsible for some 500 individual growers in total. Not surprisingly, therefore, the capacity of VEPA to exert a substantial regulatory influence is extremely limited. As one industry respondent noted: 'We are far more likely to see a WorkCover inspector than anyone from the EPA.'

However, although the impact of existing regulations had not been a major factor to date, industry representatives did point to inappropriate environmental regulations being a substantial potential threat to the industry in the future. A recent report on the issue of spray-drift was highlighted as a prime illustration of this kind of threat. Indeed, one industry respondent suggested that:

> If this [report] were to be adopted it would put half the industry out of business. The problem is that the recommendations [in the report] say that you can't spray within a hundred metres of your boundary. My whole property is only a hundred metres wide. If the industry doesn't respond to this report, it may be implemented. This is one reason why we need to take the initiative.

---

6   Note also the establishment of the European Retailer Produce Working Group (EUROP) guidelines for good agricultural produce, which seem likely to become the industry standard against which national assurance schemes will be benchmarked.

The second broad question raised above, 'why choose a partnership with VEPA?', has caused considerable anxiety among the industry because a partnership with VEPA is perceived to bring considerable risks. In particular, it is feared that this partnership will involve 'giving away some of the control over the project', that VEPA will exert excessive influence over the process, and that signing a contract with VEPA has unnecessarily aligned the industry with VEPA's agenda. Some also believe there is a risk that the involvement of VEPA will lead to regulatory sanctions being imposed on vegetable growers through greater inspectoral attention. Given these concerns, why has the industry supported the involvement of VEPA? Why not choose to proceed with an industry environmental improvement plan with far less direct input from VEPA?[7]

In essence, the answer is to provide credibility. The industry fears that, if it were to proceed with the implementation of an environmental management system *in the absence of substantial VEPA engagement*, then others, including the supermarkets and the wider community, might not believe its claims to be achieving sufficiently high environmental standards. The industry judged the benefits of VEPA-derived credibility to outweigh any risks that such a partnership might bring. Significantly, the funding that VEPA has committed to the project was rejected explicitly as an enticement. As one industry representative noted:

> We don't need or want the money. In fact, it has been more of a burden than a
> gain. Every cent of their money we spend has to be tied down in the contract—
> an arrangement that fits with their priorities more than ours.

Related to a (qualified) desire for VEPA involvement was a determination that the industry should avoid the major mistakes that occurred with the introduction of quality assurance. In particular, the industry was anxious to avoid being placed in a subservient, reactionary position relative to the major supermarket chains. There was consensus among all industry respondents that the quality assurance system now in place in the industry was inappropriate, administratively complex and costly. In short, the supermarkets had imposed quality assurance, and it had not been tailored to the needs and circumstances of the industry. It should be noted, however, that it was not the supermarkets that were blamed for this unfortunate state of affairs. On the contrary, the industry respondents were quite adamant that it was the industry itself that was to blame: despite plenty of warning that quality assurance was an inevitability, the vegetable growers had chosen to ignore it until the supermarkets had no choice but to take action. As one respondent commented: 'We all sat around with our heads in the sand. Some even formed a pact to resist its [quality assurance] introduction.'

This negative experience with the implementation of quality assurance standards has clearly galvanised the industry, in the case of the environment, to avoid repeating the mistakes of the past and to take the initiative in a co-ordinated and co-operative fashion from within its own ranks. The involvement of VEPA fits in with this strategy, and it is anticipated that it will assist in gaining the support of the supermarkets and thus obviate the need for an external (and potentially inappropriate) set of standards to be imposed.

Notably absent from the list of industry motivations was the possibility of generating a financial gain. Briefly, the proponents of cleaner production often argue that it can

---

7   Since an Environment Improvement Plan is a legislative instrument overseen by VEPA, some
    degree of direct government involvement in its implementation is inevitable.

contribute to a business's 'bottom line' through increases in production efficiency, access to new markets, or the ability to charge a price premium. None of the industry respondents, however, believed this to be a serious motivator. On the contrary, the expectation was that the adoption of environmental management systems was going to *cost* money, not *save* money. While the potential for future financial gains from industry participation in this initiative cannot be ruled out, it is clear that the pursuit of such gains has not been a significant factor in securing their introduction. Rather, the overwhelming motivation was to improve both the image and the reality of the industry's environmental performance, in the face of public pressure, potential consumer demand and anticipated supply chain pressure. As one industry respondent stated: 'We have got to go down the EMS route . . . we have got to demonstrate to the public that we are serious about tackling environmental issues.'

It is also important to examine the motivation for entering this environmental partnership from the other side, and to ask: what was the attraction of a cleaner-production partnership to VEPA? And, further, what can it bring to the partnership? In other words, what are the reciprocal benefits and obligations from VEPA engagement?

One obvious benefit to VEPA, from a broader public perspective, is that through a cleaner production partnership the vegetable growers may improve their environmental performance to a level that is greater than that achievable through traditional regulatory approaches alone. There will be additional benefits if this improvement can be made self-sustaining and if it increases progressively over time. It is also hoped that the industry's efforts will act as an example for other sectors to aspire to and, hopefully, emulate.

As noted above, in many ways VEPA (not unlike many other progressive regulatory bodies) is attempting to reinvent itself so that it is no longer solely dependent on traditional regulatory approaches. Instead, the emphasis is on getting industry to assume a larger share of the regulatory burden through various co-regulatory and/or self-regulatory arrangements. The reasons for this are threefold: first, that a greater sense of ownership on the part of industry will lead to improved environmental outcomes; second, that tailored and flexible industry strategies will reduce compliance costs; and, third, that VEPA will also be able to reduce its enforcement costs (and therefore redirect inspectoral resources to more recalcitrant industries and/or to other co-operative initiatives).

As part of this 'reinvention' strategy, VEPA has been extolling the virtues of cleaner production for many years (again, not unlike many other progressive environmental regulators). However, the evidence is far from convincing that, apart from a relatively small number of isolated examples, the uptake of cleaner production across entire industry sectors has been successful, particularly in the case of small business. In this regard, existing and previous policy approaches of cleaner production grants and case studies may be judged to have fallen below initial expectations. For example, Hamilton *et al.* (1999) noted that 'to date only a relatively small proportion of industry has experienced the benefits of a cleaner production approach. In particular, broad uptake of cleaner production is yet to occur in small to medium-sized businesses.'

It is in this context that VEPA has turned to the 'partnership' model as a potential means of salvation. However, this policy approach is still in its infancy, and there are only a relatively small number of industry participants. The participation of the vegetable growers (along with the wine industry and vehicle body shop repairers) is an important test of its efficacy. It is also consistent with VEPA's shift away from addressing larger point-sources of pollution to addressing smaller, more diffuse sources.

In contrast to the views of the vegetable growers, when a VEPA official was asked to nominate VEPA's major contribution to the cleaner-production partnership, the response was couched in purely monetary terms. That is, VEPA was contributing A$50,000 to partially fund the project. Beyond this, few other VEPA 'contributions' to the partnership were acknowledged.

Future challenges

Having described the key drivers of both industry and VEPA involvement in the cleaner-production partnership, we turn our attention now to the policy challenges that lie ahead. While some of these are industry-specific, others will have a policy resonance beyond the confines of the vegetable growers.

VEPA's role

As is clear from the earlier discussion, some vegetable growers are ambivalent, or least hesitant, about the involvement of VEPA as a cleaner production partner. On the one hand, the industry desires the added credibility that VEPA can provide, and recognises the importance of this credibility in both the eyes of the community and commercial markets (the two are of course linked, as it is the community from which final consumers are drawn). On the other hand, the industry fears ceding too much control to VEPA and, consequently, being forced to 'dance to their tune', with the possibility of divergent aims and ambitions. Part of the problem may be traced back to the fact that, beyond providing initial seed funding, VEPA has not articulated a broad range of benefits for the industry participants that it can bring to the table.

In order to overcome such lingering doubts on the part of industry (whether justified or not), and to ensure that this partnership achieves its environmental goals, it may be necessary for VEPA to develop and offer a suite of policy options designed to assist vegetable growers in achieving the partnership objectives. However, mindful that VEPA resources are extremely limited, these resources must be carefully focused on a small number of strategies capable of providing the 'biggest bang for the regulatory buck'.

First, at the end of the pilot project it is intended to develop a set of environmental management guidelines on what good management practice might involve. Such guidelines would have considerably greater credibility if they are developed not only transparently but also with VEPA and community involvement. Consideration should be given at a later stage to the possibility that such guidelines should subsequently become an industry code of practice (compliance with which would more than satisfy all regulatory requirements). VEPA should ensure that the guidelines facilitate identification and control of risks and are developed in a manner that would enable their use as a self-audit tool. This has implications for enforcement, which would be confined to those who do not take advantage of self-audits (see below). While dissemination of the guidelines to industry association members is likely to be more effective and credible if undertaken by the industry association itself, there may still be an important role for VEPA to ensure that non-members (currently some 75% of all growers) are aware of the guidelines.

Second, incentives will be necessary to ensure that a substantial proportion of vegetable growers actually adopt the guidelines, particularly since, as indicated above, the large majority of growers see no financial advantage in doing so. Ideally, this might involve subsidies, the provision of technical advice via extension services, and assistance in conducting audits. More realistically, given serious resource constraints, it might involve positive publicity to promote industry achievements or the use of cleaner production awards and green logo recognition for those who adopt formal and independently accredited environmental management systems. Some of these initiatives might also assist in the generation of financial rewards for industry participants—a motivator that was notably absent to date from the experience of vegetable growers.

Third, voluntary measures and incentives work best when they are underpinned by credible regulation and enforcement for those who are unwilling to participate voluntarily. However, with only one regulator assigned, part-time, to the vegetable-growing industry, and little capacity to allocate additional resources, the prospects for direct regulatory intervention look bleak. Yet, even with these resource constraints, there are opportunities, through the use of administrative notices and on-the-spot fines, to generate the perception of a credible regulatory presence as part of a broader policy mix, capable of dealing with both leaders and laggards.

A simplified EMS

Like many SME sectors, the vegetable growers do not have a reputation for administrative sophistication and diligent record-keeping. There is a risk, therefore, that the introduction of an environmental management system, at best, will become a resented exercise lacking in any real commitment and, at worst, totally ignored, if it is not tailored specifically to the needs and circumstances of the average or even below-average grower. A related risk is that the environmental guidelines will aim too high, and be beyond the reach of many vegetable growers. These problems may be compounded by an ever-increasing range of administrative obligations on small businesses, from food and worker safety requirements to quarterly goods and services tax returns.

A potential solution to this problem would be for the VVGA to introduce environmental management guidelines that are highly practically oriented, avoid excessive paperwork and target good rather than necessarily best practice. This could be achieved, without holding back the legitimate aspirations of committed industry leaders, with a two-tiered EMS approach. The first tier, targeted at the vast majority of growers, would be a very simple set of core management practices. There would be minimal paperwork, and the system should be self-reinforcing based on a simple 'checklist' approach. That is, rather than requiring comprehensive record-keeping, owner-operators would only be required to tick a series of boxes on the completion of each core activity (in effect, a simplified self-audit). The second tier, targeting more sophisticated operations, would be a more conventional EMS approach. This could, for example, be based on ISO 14001 and be subject to external accreditation to provide greater credibility. Over time, preferably on a voluntary basis, vegetable growers could progress from the first to the second tier.

For this approach to yield the best results, however, there would need to be a reason for vegetable growers to aspire to the more sophisticated EMS model. As indicated above, one incentive might be an environmental logo reserved exclusively for those in the second tier. A second incentive might be the provision of additional training for those vegetable growers wishing to make the transition. A third, and more interventionist, step would be the imposition of a defined time limit for tier-one vegetable growers—for example, they might be given three years to use the simpler environmental management guidelines, after which their continued participation in the programme would be contingent on them adopting the more formal, second-tier EMS.

A further strategy, particularly relevant to the issue of ever-increasing obligations, would be to integrate the environmental management guidelines with quality assurance systems. This would have the dual benefit of reducing excessive paperwork, and improving the uptake by working from a point of familiarity. The work of Ag West in its efforts to modify and expand its SQF 2000 may be pertinent in this regard. As supermarkets position themselves to meet increasing demands for clean and green produce, both they and the growers may come to see integration of quality assurance (QA) and EMS as both effective and efficient.

Leaders and laggards

One of the crucial problems confronting the programme, and the VVGA in particular, having developed a set of best-practice environmental management guidelines, is how to extend their adoption beyond a relatively small band of committed industry leaders. Not surprisingly, such leaders tend to be the larger, more sophisticated vegetable-grower operations. A further handicap is that the VVGA membership comprises only about 25% of the total vegetable-grower industry. Consequently, its normal avenues for distributing information will clearly not reach the entire industry sector. The danger is that, without some additional policy mechanism, the VVGA will have difficulty in expanding the programme beyond a select few industry leaders. In short, it may be 'preaching to the converted', or at least to those willing to be converted. Arguably, it is those that fall outside the leadership group, the laggards or merely average, that are most in need of, and/or have most to gain from, better environmental management practices.

A judicious combination of carrots and sticks is likely to achieve the best results. Those who are willing to undertake the proposed self-audit and send the results to the regulator will be a low inspectoral priority, which means in practical terms that they will not be subject to routine inspection. Those who go one step further to adopt a formal, albeit simplified, EMS will be given a logo and public relations benefits and will also, over time, gain an commercial advantage to the extent that larger wholesalers, supermarkets, or even consumers who recognise the logo, give them preference.

On the other hand, those who take neither of these steps will receive notification from VEPA that, if their environmental practices are found wanting, they will be subject to formal regulatory action. Politically, the industry association is unlikely to object to this, since it will be largely non-members that fall within this group. In practical terms, the threat of enforcement action may be largely a bluff, but it will only take a small number of on-the-spot fines and administrative notices, appropriately publicised, to create the impression of a significant regulatory presence. And, while VEPA lacks the capacity to conduct inspections on any regular basis, it can rely on complaints from an increasingly aware and proximate urban community, as the basis for occasional and targeted inspections.

Harnessing supply chain pressure

One mechanism for greatly extending the adoption of environmental management systems across the entire vegetable-grower industry might be to employ a third-party regulatory surrogate in the form of supply chain pressure. The purpose or role of such a regulatory surrogate is to exert pressure on an industry to improve its environmental performance in addition to, or in the absence of, conventional governmental regulatory pressure. In this case, the obvious candidate is the supermarket industry. It has the necessary leverage, in the form of commercial power and a close working relationship with the industry, as demonstrated in the area of quality assurance and food safety, to greatly enhance the ubiquity of environmental management systems among vegetable growers.[8] Arguably, then, the VVGA should seek to engage the supermarket industry at some point in the pilot phase, to garner both its interest and support, and to explore options for the supermarkets to take an active role in facilitating its widespread intro-duction. Attempts to integrate environmental management guidelines with systems of quality assurance, as described above, may be of assistance in this regard.

There would be mutual gains for the VVGA and the supermarkets in such a strategy. From the industry association perspective, pressure for an EMS or other form of environmental assurance from the supermarket industry is very likely in the medium to longer term and it is far better to take the initiative and become an active partner in this process rather than risk, as in the case of quality assurance, standards being imposed on it. From the perspective of the supermarkets, it is apparent from the international con-sumer experience that demand for 'clean and green' produce is increasing, and that those who position themselves to demonstrably supply it will gain a competitive advantage.

In addition to a regulatory surrogate 'stick', a supplementary 'carrot' to widespread participation would be to focus on the financial benefits that could accrue to individual vegetable growers. In this regard, noting the aforementioned degree of scepticism evi-dent in the industry, the lure of internal financial efficiency gains from cleaner production practices is unlikely to prove sufficient. A more powerful signal might be the use of an environmental logo, conferred by the VVGA and endorsed by VEPA, enabling vegetable growers to capture a greater share of existing markets, access to new markets and/or a price premium. The rapid growth in demand for organic fruit and vegetables is instruc-tive in this instance (Moynihan 2000). The use of logos would obviously need to be co-ordinated with its use to achieve other policy ends, such as the earlier suggestion for it to act as an inducement for growers to progress from streamlined to more compre-hensive EMSs.

Community acceptance and participation

As described above, arguably the key driver of industry moves towards the introduction of best-practice environmental management guidelines has been community pressure. A key concern for the vegetable-grower industry, therefore, is community acceptance of its efforts. In part, the credibility of growers' efforts may be assisted by VEPA involvement as cleaner-production partners (external accreditation, referred to above, is another means of increasing community acceptance).

---

8 For example, in the UK, the Tesco supermarket chain runs its own assurance scheme, Nature's Choice, which 'includes guidelines on chemical inputs, energy and water efficiency, worker health and safety, wildlife and landscape conservation. At least 95% of Tesco's fresh produce sourced in the UK is now certified under the scheme and the standards are being introduced to suppliers worldwide' (*Tomorrow* magazine, September/October 2000: 23).

Beyond VEPA's role, greater community acceptance could be achieved by having broader stakeholder representation in the management of the industry's EMS programme. As noted above, the industry has established a steering committee to oversee the introduction of best-practice environmental management guidelines. Conspicuous by its absence, however, is any community representation. Inviting participation from, for example, a member of the Western Point Action Plus group may have positive outcomes for the credibility and transparency of industry efforts. Some in industry may be nervous at such a prospect, and there is the possibility that such representation will hamper progress. Despite these potential shortcomings, the potential benefits of broader stakeholder representation are great enough to warrant its serious consideration.

From a VEPA perspective, the community can play a number of other important roles. For example, given the inability of VEPA to inspect vegetable growers to identify serious breaches, community complaints may be particularly valuable in identifying egregious breaches. A formal complaints line, well publicised, would be useful to facilitate such complaints. It is also important that the whole process of developing environmental guidelines, their terms, and the conditions for the award of logos, are all transparent. The fact that the entire process is being developed through the mechanism of an EIP (see Chapter 8), whereby community involvement is an important component, should ensure that this is indeed the case.

Industry co-operation

A recurrent theme among several industry respondents was the apparent lack of a 'culture of co-operation' within the vegetable-grower industry. Reasons put forward for this include the lack of a formal industry co-operative structure that is often associated with single-commodity agricultural sectors, and a tendency for vegetable growers to undercut one another in the marketplace. It is possible that this lack of co-operation will hamper VVGA efforts to introduce best-practice environmental management guidelines across the industry.

Rather than be handicapped by a lack of co-operation, the industry may be able to 'turn the tables' by using the introduction of best-practice environment management guidelines as a means of building industry camaraderie and raising its overall level of professionalism. It may also be possible for the VVGA to leverage off the guidelines to increase its membership base. A pertinent example of how environmental improvement can be used as a catalyst and means of achieving broader industry objectives is provided by the air-conditioning and refrigeration sector of Western Australia outlined in Chapter 4. In this case, the advent of obligations to reduce the use and release of ozone-depleting gases saw the industry establish an industry association, develop a united position for the negotiation and implementation of a self-regulatory scheme, introduce codes of practice, and institute minimum standards of training. The net effect of these developments, beyond the obvious environmental benefits, was an industry that is far more organised, coherent and focused in its policies and practices than was previously the case.

Conclusion

The vegetable growers of Victoria provide an interesting example of an industry that is at a crossroads: it recognises that the environmental practices of the past are no longer acceptable (in face of community and consumer demands) but is not entirely certain of the best way forward. On one hand, it is an industry that is not comfortable with, or used to, government engagement, particularly from regulatory agencies. On the other hand, it recognises that 'going it alone' would leave it exposed to future overriding action by the commercially powerful supermarket chains (which has already occurred in the case of quality assurance). It is also mindful of the limited resources and skills of its membership in seeking to adopt environmental management guidelines. This has led it, perhaps somewhat reluctantly, to accept a partnership arrangement with VEPA as a necessary means of obtaining broad community credibility.

It is apparent, however, that the industry is not yet contemplating a 'deeper' partnership with VEPA, or indeed other stakeholders, beyond that of financial input. There may be several unexplored opportunities to extend both the coverage of environmental management guidelines and their efficacy in generating improved economic outcomes. For its part, VEPA is attracted to a cleaner production partnership with the vegetable growers because it closely matches its increasing focus on small business and non-point sources of pollution, and its emphasis on innovative, non-regulatory policy approaches. However, it appears reluctant to provide additional incentives and/or benefits to industry participants in the programme. This mismatch of needs, expectations and provisions, while not likely to threaten the existence of the partnership programme, does have the potential to undermine its overall effectiveness.

If the vegetable growers' partnership is to prosper rather than merely survive, then further measures will be necessary. Not least, VEPA must be prepared to contribute not just financially but also in substance; efforts must be made to harness the supply chain pressure that the supermarkets can provide; and the community must be empowered to become an active participant in the process and outcomes of achieving cleaner production. It is only by harnessing these pressures, by nurturing cleaner-production initiatives through an industry partnership, and by providing an underpinning of regulation for worst performers who do not respond, that regulators can achieve a long-term improvement in industry standards.

6
REGULATING
LARGE ENTERPRISES

The task of regulating large enterprises is very different from that of controlling the behaviour of SMEs. Most such enterprises, unlike SMEs, are sophisticated in their general (and increasingly in their environmental) management, and have substantial units devoted to legal and environmental issues. They commonly have long-term business plans, complex systems of controls designed to manage business and legal risks, including EMSs, and, equally important, deep pockets.

In most industry sectors, some large laggards remain, but an increasing number of large enterprises are also increasingly accepting of their obligation to comply with environmental regulations (although they often lobby fiercely to block new ones) (see e.g. Howes *et al.* 1997). Some, indeed, have set environmental goals intended to take them substantially beyond compliance with existing environmental regulations, and others build in a 'margin of error' which ensures that, even when unanticipated fugitive emissions occur, they remain within their legally prescribed emission limits (Gunningham and Sinclair 2002).

A minority have shifted their thinking even further, to the extent that they now believe there are opportunities to achieve win–win outcomes, whereby increased environmental expenditure, far from representing a business cost to be avoided if at all possible, represents an opportunity for economic gain. 3M's Pollution Prevention Pays programme is the best-known example of this approach, although it must be emphasised that such approaches remain the exception rather than the rule, and the majority still view with suspicion the idea that there may be a 'free lunch' in environmental protection.

Many large enterprises are also reputation-sensitive. They have become acutely aware that damage to their corporate image caused by poor environmental performance can undermine their share value, prejudice their standing with governments and thereby threaten investment opportunities, bring unwanted attention from regulators, and incur the wrath of environmental groups and local communities. All of this, as we will see, can seriously threaten a firm's 'social capital' with a range of economically damaging consequences, and makes such companies vulnerable to 'corporate shaming' in a way that most SMEs are not.

Large enterprises are elephants rather than foxes: they are big, readily identifiable and easy to target. Whereas regulators may have considerable difficulty even finding some SMEs, and can inspect them only very rarely, this is not the case with large enterprises. The latter are commonly major point-source polluters, and as such are subject to a series of detailed obligations under the terms of plant-specific licences or permits. Commonly such obligations include continuous monitoring of certain pollutants, and a duty to self-report any breaches of their licence requirements. Accordingly, large enterprises can anticipate regular inspections from regulatory agencies. They also commonly attract the attention of, and scrutiny and pressure from, environmental organisations and local communities.

These characteristics facilitate the use of a range of regulatory strategies that may have little resonance for SMEs, and make redundant the use of others that are fundamental to successful SME regulation. For example, education and compliance assistance strategies, while crucially important in relation to SMEs, have little, if any, application to large enterprises. Indeed, the latter often know far more about the environmental problems they confront, and have access to far greater legal resources, than do regulators themselves. In contrast, other strategies, such as 'shaming' poor environmental performers through informational regulation, while almost entirely irrelevant in the case of most SMEs, can be particularly potent when applied to large reputation-sensitive organisations.

Again, because the population of large enterprises includes those that are contemplating going beyond compliance, policy instruments directed at this group should be designed not merely to drag industry laggards up to the standards of minimum compliance but also to reward, facilitate and encourage environmental best practice and environmental leadership. Regulatory flexibility initiatives are particularly apposite in this regard.

Finally, large enterprises, by virtue of their often sophisticated systems of internal control (including EMSs), have a much greater potential capacity for environmental self-management and self-monitoring than their smaller counterparts, although whether they have a self-interest in developing it is another matter. Sometimes this capacity can be harnessed by the relevant industry association, in circumstances where the industry as a whole has a collective interest in improved environmental performance (for example, when its legitimacy is under threat and it contemplates a more draconian alternative). The limited number of players in some industries, their high visibility and interest in participating in the relevant industry association, and the consequent capacity to curb free-riding, are further reasons why various forms of self-regulation, although problematic, are perceived to be more credible in the case of large businesses than in SMEs.

For these reasons, we do not propose to examine the entire range of policy instruments in this chapter but rather focus on those that are particularly apposite to the circumstances of large enterprises identified above. Rather than dwelling on conventional or first-generation policy instruments which have already been examined fully by ourselves and others (Gunningham and Grabosky 1998; Hutter 1996, 1999), we focus on successful innovations, and on what have been described as next-generation environmental instruments and strategies (Chertow and Esty 1997; Center for Strategic and International Studies 1998).

The remainder of this chapter is in five sections. First, we examine the fastest-growing policy instrument of the last decade: voluntary approaches. Second, we examine various 'regulatory flexibility' initiatives, which provide an alternative regulatory track, at least for environmental leaders. Third, we consider economic instruments, focusing on the three instruments most commonly employed to deal with issues of point-source pollution: taxes or charges, tradable permits, and subsidies. Fourth, we explore the role of informational regulation, a relatively new but potentially cost-efficient and effective means of achieving environmental improvement in reputation-sensitive organisations. Finally, we explore the implications for enforcement of the new generation of environmental policy instruments and how to achieve best practice in terms of environmental compliance. Here we examine the role of third parties as potential 'surrogate regulators'.

By the end of the chapter we hope to have provided a critical evaluation of next-generation instruments designed not only to address laggards but also to encourage environmental best practice and to relate this evaluation to broader developments in regulatory design internationally. In the following chapters we complement this with our own case studies to demonstrate how at least some of these initiatives can best be implemented in practice, the problems they may confront, how these might be overcome, and how tangible improvements can best be achieved.

Voluntary approaches

Voluntary approaches include self-regulation, voluntary codes, environmental charters, co-regulation, covenants and negotiated environmental agreements. Over the past few years such approaches have become an increasingly popular environment protection tool, and, in one form or another, their use has spread worldwide. Negotiated agreements are particularly popular in the European Union, where there are believed to be over 300 such agreements, almost all focused on large enterprises. While in principle this approach could be extended to SMEs (and we explored one such example in Chapter 4) this is unusual, at least in the European and North American contexts.[1]

The reasons for this new-found interest in voluntarism are many (Moffet and Bregha 1999) but include the limitations of command-and-control regulation, the need to fill the vacuum left by the retreat of the regulatory state, and the interest of industry itself in seeking (at best) a flexible, cost-effective and more autonomous alternative to direct regulation, or (at worst) a means of avoiding the imposition of binding standards altogether. This development also reflects, as the European Union's Fifth Action Plan points out: 'the growing realisation in industry and in the business world that not only is industry a significant part of the (environmental) problem but it must also be part of the solution' (CEC 1996: 561). Voluntarism, as a result, places substantial responsibility on industry itself, to design its own environmental policy and the means by which it will be

---

1 Large organisations are not only better organised and often more proactive; they also commonly have well-developed industry associations catering for their needs and capable of negotiating on their behalf. For these reasons large enterprises have been better able to work strategically in their perceived long-term self-interest to negotiate such agreements, often as a more palatable alternative to legislation. However, it should be noted that there are about 30,000 local pollution control agreements in Japan, including agreements with SMEs. See OECD 1999.

implemented, on the assumption that industry knows best how to abate its own environmental problems.

From an environmental policy perspective, the increasing reliance on voluntary approaches raises a number of important issues. Not least, how do they work, where do they work, what are their strengths and limitations and how can they best be used within the overall framework of environmental policy design? It will be particularly important to ascertain how well such approaches work compared to the available alternatives, how efficient or effective they are, to what extent they can be relied on as a substitute for other policy instruments and the extent and the circumstances in which they can be better used in complementary combinations. In the following sections, following the classification developed by the OECD (1999b), we explore these questions in relation to the three types of voluntary arrangement which are the most prevalent and the most important in environmental policy terms:

- **Unilateral commitments**, set by the industry acting independently without any involvement of a public authority

- **Public voluntary programmes**, which involve commitments devised by the environmental agency and in which individual firms are invited to participate. Participation in the voluntary programme is a choice left to individual companies.

- **Negotiated agreements**, which involve commitments for environmental protection developed through bargaining between a public authority and industry. They are frequently signed at the national level between an industry sector and a public authority.[2]

As we will see, the term 'voluntary' itself, although almost universally used to describe the above types of arrangement, may be a misnomer, in that effective initiatives are rarely, if ever, *purely* voluntary in nature. On the contrary, firms are usually pressured into entering into such arrangements, either by threats of legislation or by positive inducements such as tax relief or subsidies, or by the broader expectations of civil society in general and NGOs in particular.

Unilateral commitments

Unilateral commitments consist of environmental improvement programmes set up by enterprises or industry associations and communicated to their stakeholders, with both the targets and determinations of how they are to be met and monitored at the discretion of the enterprises or associations themselves. Although such commitments do not involve other groups directly, a major motivation for entering into them is to protect corporate or sector environmental reputation by gaining credibility with stakeholder groups. This in turn may serve to forestall anticipated regulation, ease the compliance burden, mitigate stakeholder pressure for improved environmental performance, or earn rewards from customers, financial markets or consumers.

---

2 Consistent with the OECD approach, we do not explore private agreements reached through direct bargaining between stakeholders, such as polluters and pollutees, because very little information is available concerning such agreements.

To date, unilateral commitments have taken two main forms. The first and most common form of commitment by individual enterprises is to sign up to a code of conduct or environmental charter, which seeks to encourage higher standards of corporate environmental performance across industry as a whole (but does not involve a collective or sector-specific performance target). Such codes are designed to influence behaviour within enterprises and to influence outside perceptions of this behaviour. Potentially, they can provide a moral compass and a systematic means of improving compliance and achieving broader environmental aspirations, act as a communications tool and promote a culture of environmental integrity (OECD 2000b).

An example of this approach is the principles developed (for adoption by individual enterprises) by the Coalition of Environmentally Responsible Economies (CERES), which brings together a number of major US environment groups and various socially responsible investors and public pension funds.[3] These ten principles cover the protection of the biosphere, sustainable use of resources, disposal of wastes, energy conservation, risk reduction, safe products and services and environmental restoration, issues of public information, management commitment, and audits and reports. There is considerable emphasis on monitoring, implementation and reporting on progress. A number of leading and reputation-sensitive North American and European corporations have adopted such codes.[4]

The second form of individual commitments are those developed by an industry association which aspire to apply to the entire industry sector (i.e. industry self-regulation). By far the most significant example of this approach is the chemical industry's Responsible Care initiative, which operates in over 40 countries. While different countries have implemented Responsible Care in different ways, at its core it includes commitments to improved environmental performance (though without specifying measurable outcomes), to improved relations with customers and communities, and to greater transparency. These aspirations are to be achieved through a series of codes of practice, many of which relate to good management practices and systems (and which provide both a public commitment to improvement and a 'moral compass'), and through mechanisms that seek to develop greater dialogue with the local community. Responsible Care is discussed in more detail in relation to the mining industry in Chapter 7.

These types of unilateral voluntary commitment are not public policy instruments per se because they are taken exclusively by private-sector organisations. Accordingly, while governments may encourage such initiatives—for example, by recognising and publicising them—they cannot use them as a public means to achieve environmental goals, However, unilateral agreements do have public policy implications, and governments can

---

3   Also of future significance may be the revised set of OECD Guidelines for Multinational Enterprises (MNEs). These are non-binding recommendations to enterprises, whose aim is to help MNEs operate in harmony with government policies and societal expectations, by providing guidance on appropriate business conduct across the full range of MNE activities. Although the environmental issues that have been built in are still relatively modest, the Guidelines may ultimately come to 'serve as an independent benchmark of the state-of-the art thereby helping to harmonise objectives among government, business, labour and other stakeholders' (Johnstone 1999: 18).

4   For example, in addition to participation in CERES, approximately 2,000 companies have signed the International Chamber of Commerce's Business Charter for Sustainable Development.

encourage such activities, albeit indirectly (OECD 2000c). It is important therefore to assess their value and the circumstances in which they are most likely to function in the public interest. Despite the paucity of evaluative research,[5] a number of general conclusions can be drawn about the value of unilateral agreements and their potential role as an environmental policy instrument.

First, unilateral commitments at industry level (i.e. an industry-based code of practice under which individual enterprises set their own specific individual targets) are likely to work best when the following conditions are present: there are relatively few industry players; the exit costs are high (for example, quitting the scheme will draw adverse reaction from markets, competitors or regulators); there is a history of co-operation between member companies; expertise and resources for regulation are available in the industry; non-compliant behaviour can be punished; consumers value compliance; fair dispute settlement mechanisms are in place; and some role is available for public participation or oversight (Priest 1999–98).

Second, a major concern for such collective initiatives is to curb the incidence of free-riding, whereby rogue firms seek to claim the public relations and other benefits of scheme membership while avoiding the obligations it entails. Unfortunately, free-riding is often an almost insurmountable problem, because the criteria identified above, or any approximation to them, are only likely to be met in a small number of circumstances. In practice, individual targets are often set to a lowest-common-denominator level and are not measurable, enforcement is often (but not invariably)[6] weak, and such initiatives commonly lack many of the virtues of conventional state regulation, 'in terms of visibility, credibility, accountability, compulsory application to all, greater likelihood of rigorous standards being developed, cost spreading, and availability of a range of sanctions' (Webb and Morrison 1996).

Third, the paucity of success stories in the empirical literature[7] should make governments extremely reticent about relying on unilateral programmes as a basis for providing any form of regulatory relief or other concessions, notwithstanding industry suggestions that it should do so. Indeed, the history of the Institute of Nuclear Power Operations (INPO), which is arguably the most successful such initiative of all (see Box 4), suggests that an underpinning of government regulation and enforcement is crucial in maintaining credibility and effectiveness.

From the above, it may be safely concluded that 'pure' unilateral commitments at industry group level cannot be relied on to deliver public policy outcomes, even in the most favourable of circumstances. There is even less available evidence concerning the degree to which, or the circumstances in which, individual commitments by individual enterprises are effective. Nevertheless, it seems reasonable to extrapolate from the experience of industry-level commitments to draw a similar (albeit provisional) conclusion about individual voluntary commitments, at least in the absence of evidence to the contrary.

Such commitments may provide a number of 'soft effects'. For example, unilateral commitments by individual enterprises in terms of adopting generic codes may result in

5 For recent studies, see King and Lenox 2000; Nash 2000; Gunningham and Rees 1997.
6 It should be noted that the American Forest and Paper Association had, by mid-2001, expelled 16 industry members and that contract-based enforcement was increasingly being viewed as a credible substitute for self-regulators' lack of criminal enforcement powers.
7 See generally OECD 1999, especially Chapter 3 and references therein.

Background and compliance problem

Early in the morning of 28 March 1979, several water pumps stopped working at a nuclear plant near Harrisburg, Pennsylvania, setting in motion the chain of events that would make Three Mile Island (TMI) an indelible symbol of disaster for the public. The disaster not only cast doubt on the credibility of the existing system of government regulation, but also placed a cloud over the future of the US nuclear power industry. Another such disaster could easily lead to a public and regulatory backlash capable of threatening the very viability of the industry itself. How could the industry prevent 'rogue' operators putting the entire industry at risk?

The compliance innovation

Less than two weeks after the accident, a committee of nuclear utility chief executive officers gathered to co-ordinate the industry's response to the accident, and the efforts resulted in perhaps the single most important change that has taken place in the post-TMI nuclear regulatory regime. They created a private regulatory bureaucracy—the Institute of Nuclear Power Operations (INPO)—and today (with an annual budget of nearly US$54 million) INPO's approximately 400 employees develop standards, conduct inspections and investigate accidents.

The results

By all accounts, the safety of nuclear plants has increased significantly since TMI, and there is wide agreement among knowledgeable observers (proponents and critics of nuclear power alike) that INPO's contribution to improved nuclear safety has been highly significant (Rees 1994). This has been achieved by education, persuasion, peer-group pressure, gradual nagging from INPO, shaming, or the other instruments of informal social control at INPO's disposal.

However, even INPO is incapable of working effectively in isolation. There are, inevitably, industry laggards who do not respond to such informal pressure. After some years of achieving very little progress in changing the behaviour of this minority, INPO faced a dilemma as a significant number of plants were ignoring the problems INPO had identified, yet getting tough seemed out of the question because that might drive the recalcitrants out of the association. INPO's ultimate response, after five years of frustration, was to turn to the government regulator, the Nuclear Regulatory Commission (NRC).*

The result was the effective dismissal of top executives, plant shutdown and substantial improvements in safety. All this was achieved simply by invoking the power of the NRC (which alone had the capacity to bring criminal proceedings and to shut the plant down). Had effective action not been taken against the recalcitrant few, and, in the longer term, had free-riders been allowed to flourish without sanction, then INPO's authority over other firms would have been jeopardised. As Rees notes, INPO's climb to power has been accomplished on the shoulders of the NRC.

* For example, Rees (1994) describes how, in one case, INPO sent a letter to the CEO of one company (12 pages plus 17 attachments) documenting its persistent failure to address a number of very serious safety concerns, and insisting that the CEO pass a copy of the letter to his own board and the NRC.

Box 4 *Institute of Nuclear Power Operations*

Source: Rees 1994

the accumulation of managerial expertise in ethical and legal compliance. According to the OECD:

> voluntary initiatives in corporate responsibility have promoted the accumulation of management expertise needed to translate legal compliance and ethical norms into the day-to-day operations of companies. The institutional supports for this expertise—management standards, professional societies, specialised consulting and auditing services—help lower the cost of legal and ethical compliance as well as make it more effective (OECD 2000b: 3).

In relation to industry-level unilateral commitments, these soft effects may also be important, although they are often ignored.[8] For example, according to Rees (1997) Responsible Care has the capacity to build an industry morality: a set of industrial principles and practices that defines right conduct and spells out the industry's public commitment to moral restraint and aspiration. It is also capable of institutionalising responsibility through the development of industry-wide policies and procedures to ensure a strong and effective commitment to the values or ideals the industry claims to uphold, the integration of accountability and transparency in corporate decision-making, and the capacity to 'moralise social control' (Rees 1997).

In summary, the debate is increasingly about how, and in what circumstances, it may be possible to design effective co-regulation or industry self-management.[9] Relevant variables include the degree of coincidence between the self-interest of the individual company or industry, and the wider public interest, and the degree to which enterprises and/or other third parties (such as suppliers, buyers and environmental groups) can detect non-compliance. In Chapter 7 we explore these issues further though our study of voluntary initiatives in the mining industry, from which we hope to extrapolate more general lessons on the potential role that voluntary approaches can play in improving environmental performance. In particular, we hope to emphasise the critical importance of issues of coverage and ambitiousness, monitoring, sanctions for non-compliance, transparency and credibility of voluntary approaches.

Public voluntary programmes

Public voluntary approaches have been defined as programmes devised by an environmental agency in which individual enterprises are invited to participate. In their most common form, the regulatory agency pre-sets a target and invites individual enterprises to commit to it. Inducements to join may include some technological or financial assis-

---

8 For example, most assessments of Responsible Care focus on the extent to which it succeeds (or fails) to curb the excesses of its members by external means (and on its incapacity to apply effective sanctions and to curb free-riding). In contrast, Rees's (1997) careful and sophisticated analysis of the same industry programme, described more fully in Chapter 7, focuses on how it achieves behavioural change through two crucial *internal* processes.

9 Co-regulation occurs usually where industry develops and administers a code and government provides the ability to enforce the code by giving it a legislative backing in some way. Under co-regulation government involvement usually falls short of prescribing the code in detail in legislation.

tance and public relations benefits: for example, the right to take advantage of a green logo. Enterprise participation is purely optional. Although the incidence of voluntary programmes has been increasing, and some, such as EMAS, described below, have a high profile, the overall number of such agreements is not large. In the European Union, at the time of writing, there were estimated to be less than 20 such agreements, and about double that number in the USA (Mazurek 2000).

Industry 'challenge' programmes are perhaps the best-known form of public voluntary programme. Many of these are within the areas of energy efficiency and chemical use reduction, where government hopes to stimulate enterprises to identify win–win solutions. Most of these agreements, at least in the USA, 'require participants to sign non-binding letters of agreement such as a Memorandum of Understanding (MOU) which imposes no sanction for program withdrawal. Failure to meet the MOU terms means that the company can no longer claim the benefits of participation, which usually consist of public recognition' (Mazurek 2000: 4). Two of the most notable examples are the US EPA's 33/50 and Canada's Accelerated Reduction and Elimination of Toxics (ARET) programmes.

Under 33/50, firms are encouraged to reduce the release of toxic chemicals (Miller 1994). Industry participation is completely voluntary and commitments are not enforceable by law. Instead, the programme relies on co-operation between industry and the EPA, and subsequent positive public recognition of environmental achievements. Participating enterprises pledge a reduction in their releases of any of the specified chemicals and to develop detailed action plans. Evaluations reveal that participants in the 33/50 programme have significantly reduced toxic chemical releases, although whether they would have achieved similar reductions in the absence of the programme is unclear (Arora and Cason 1995; Ransom and Lober 1999). Harrison (2001) argues that many of the reductions might have been achieved under a business-as-usual scenario.[10] However, one study suggests that, even taking account of the self-selection biases, 33/50 resulted in emissions reductions of 28% relative to pre-programme levels (Khanna and Damon 1998, cited in OECD 1998b).

Canada's ARET programme operates in a similar vein, with industry being challenged to reduce discharges of 30 specified chemicals by 90% by 2000 and 87 others by 50%. Claims of its success, like those relating to 33/50, have been challenged on the basis that the reductions achieved may or may not be voluntary and, even to the extent they are, may not be attributed to the programme (Harrison 2001). Similarly, the US General Accounting Office (1997, cited in OECD 1998b: 118) has been critical of the purported success of another US EPA Challenge initiative, the Green Lights programme, maintaining that claims that it had been responsible for influencing a quarter of all participants to adopt different technologies were greatly overstated, given that a significant part of the purported achievements were gained before the programme took off.

It may be that the greatest contribution of such programmes is in terms of soft effects that are not measurable, such as the diffusion of information on pollution abatement techniques in industry. Not least they

---

10   Some may have been driven by existing regulations, and some (predominantly reputation-sensitive large enterprises) may have been pressured to reduce emissions of 33/50 chemicals for other reasons.

commonly include technical assistance, decision-support too, best practice guidelines, evaluation tools, and training sessions, thus improving the level of knowledge in the participating companies. Besides, they frequently exhibit a function of 'signaling' via the use of a logo . . . or promotional supports. They thus improve public recognition of efforts for greening business strategy. In turn, such reputation gains provide industry with long-term incentives to commit to environmentally friendly trajectories' (OECD 1999b: 124).[11]

As with other voluntary approaches, the design of such public voluntary programmes will have a strong bearing on their potential to achieve environmental goals. The target (usually adopted at enterprise level)[12] is pre-set by government, which may be unable to distinguish between additional improvements and business as usual. It may also be tempted to lower the target to attract greater levels of participation. Or it may set goals that are very vague in order to attract participants from a diffuse range of industry types. Krarup's (2001) research revealed that not only is an ambitious target essential but it should be established in an open and transparent process (involving NGOs and parliament). Adequate data on performance and achievement will also be essential, as will credible monitoring mechanisms (for example through a third-party auditor), since without these it will be difficult to estimate whether targets have been achieved, and enterprises may be tempted to gain the public relations benefits without making the improvements. The absence of sanctions under public voluntary programmes also provides enterprises with a strong temptation to default and free-ride. The OECD has summed up their limitations as follows:

> in the absence of regulatory threat: lacking an incentive to abate above and beyond existing regulations, firms' prime motive to participate [is] either the existence of 'no regret' (profitable) pollution abatement actions or to benefit from public recognition. In general, such motives will not take firms very far in their environmental improvements, as pollution abatement would soon become a costly business (OECD 1999b: 132).

In summary, most voluntary programmes have serious weaknesses, not least of which are the lack of credible targets to take participants substantially above minimum regulatory levels (where regulation exists) or beyond 'business as usual', and the strong temptations to free-ride. While such programmes may have significant soft effects such as information diffusion and technical assistance, and can be established by government at low cost, nevertheless their overall contribution to higher environmental standards is a very modest one.

11   See also Mazurek 2000.
12   Usually the targets require implementation of particular systems (e.g. for energy efficiency, or EMS), the setting of individual firm-level goals (since in some respects each firm is unique) and the achievement of specific reductions of different kinds of emissions. As regards the latter, participating firms agree on the same targets.

Negotiated agreements

Negotiated agreements involve specific commitments to environmental protection goals elaborated through bargaining between industry and a public authority. They have been developed as part of an explicit attempt to improve environmental policy outcomes without overburdening industry or putting it at a competitive disadvantage, and in particular to promote a quicker and smoother achievement of objectives than the cumbersome and often conflict-ridden route of legislation.

In Europe they represent by far the most popular and important form of voluntary initiative. Here, they are usually entered into by an industry association and government against a backdrop of threatened legislation: the tacit bargain being that if the industry will commit to reach given environmental outcomes (e.g. an industry sector target) through its own initiatives, government will hold off on legislation it would otherwise contemplate enacting to address the problem. Most such agreements have been in areas of waste and air management, climate change, ozone depletion and water pollution. As collective agreements, they are not binding on individual enterprises, with potential sanctions being confined to the collective level.

Negotiated agreements are quite distinct from the sorts of legally binding agreements made between individual enterprises and a government agency, whereby government offers regulatory flexibility in exchange for beyond-compliance environmental performance, and whereby the individual enterprise would risk individual liability or eviction from the agreement for breach. The latter type of agreement is becoming increasingly popular in North America, and has a number of major differences to the European voluntary collectivist approach. Not least, they are integrated with, and form a part of, the conventional command-and-control regulatory system, rather than providing an alternative to it. The regulatory flexibility initiatives of the Clinton–Gore era are the clearest example of this approach. We address the issue of regulatory flexibility separately in the next section below, and confine our present examination to collective negotiated agreements made to pre-empt or render more interventionist state action unnecessary.

From a public policy perspective, the attractions of negotiated agreements include the potential to achieve environmental improvements at less cost to both government (which may avoid or reduce the costs of standard development and monitoring, but not of negotiating) and industry (by providing enterprises with implementation flexibility). It is assumed that industry will have far superior knowledge to government as to how to achieve any given outcome at least cost. However, the crucial question is whether industry will be motivated to do so under a negotiated agreement, or whether more reliable outcomes could be achieved with very much the same flexibility via other means such as the imposition of performance standards (which prescribe outcomes, not means) or economic instruments. From industry's point of view, the most important motivations for entering such agreements will relate to avoiding regulation, gaining competitive advantage, protecting or increasing sales, enhancing public image and reflecting internal values (Baeke *et al.* 1999; Stratos and Pollution Probe 2000).

There is a tension between the goals of government and industry that raises a number of challenges. First, to the extent that an agreement would commit industry to doing something it would not otherwise choose to do (i.e. spend money on environmental improvements that do not otherwise enhance profits), then the agreement must provide sufficient incentives to deliver a net gain. Such incentives might include reputation enhancement, facilitating a price premium or expansion of market share, or the provision of regulatory concessions. The last is likely to be by far the strongest incentive to join, and a substantial number of agreements have involved implicit or explicit bargains of this nature. For example, the Federated Association of German Industry in 1995 agreed to propose a reduction of carbon dioxide emissions by up to 20% by 2005, in exchange for which the federal government 'announced the withdrawal of plans to introduce a waste heat ordinance and promised an exemption from a possible energy tax' (Ramesohl and Kristof 1999).

Second, since industry would prefer to obtain whatever benefits are available under the programme at as little cost as possible, it is likely to negotiate hard to minimise its commitments. For example, under the German carbon dioxide agreement discussed above, Jochem and Eichhammer suggest that the target was so modest that 80% of German industry had already achieved it at the time it was announced (Jochem and Eichhammer 1996, cited in Covery and Lévêque 2001). The likelihood of such an outcome is increased by the asymmetry of information that characterises many government–industry interactions, with industry knowing far more about what is technologically possible and economically practicable than government.[13] Of course (in relation to the first point above) enterprises that may be able to satisfy the terms of the agreement at little or no cost (for example, targets under some energy efficiency agreements only imply 'no regrets' actions) may need fewer incentives to participate.

Third, where the costs of participation are substantial but the enterprise has sufficient incentive to join, it may seek to gain the benefits of participation without bearing the costs. A related problem arises in heterogeneous industry sectors, where some firms see reputation advantages from gaining a green image but others do not, and the latter may simply decline to enter into a collective agreement. This raises problems (depending on the design of the agreement) as to whether voluntarism really is a viable alternative to regulation for the entire industry sector. Both these scenarios are likely to play out quite frequently because, as Higley *et al.* point out: 'industries are profit driven creatures. Pollution abatement is often expensive, and the *most* cost effective option for the firm is very often to abate less' (Higley *et al.* 2001: 10).

Perversely, these tensions generate risks of a phenomenon tantamount to regulatory capture, whereby regulators acquiesce in the negotiation of targets and other conditions that are unduly favourable to industry and contrary to the public interest. Negotiated agreements are particularly fertile ground for this form of capture because, as Ayres and Braithwaite (1992) point out: 'The very conditions that foster the evolution of co-operation are also the conditions that promote the evolution of capture and indeed corruption.'

It is clear that negotiated agreements could be beset by a variety of serious problems. This raises the question of how such agreements can be best designed to overcome or at least to mitigate such problems, and whether, to what extent, and in what circumstances, existing agreements have done so, and whether they have performed better than other policy instruments. Unfortunately, the empirical literature on negotiated agreements is very limited. Many existing agreements lack clear reporting requirements and deadlines, making evaluation of their success extremely difficult. Indeed, one of the few things on which almost all analysts seem to agree is that far too little attention has so far been given

---

13   Such an asymmetry also exists, of course, in negotiating direct regulatory standards.

to evaluating either their economic or environmental benefits (Davies and Mazurek 2000; NRC 1997; Beardsley 1996; Harrison 2001).

The preliminary assessments that are available have mixed success. In some countries, and in some particular contexts, such agreements have demonstrated their value, as the Japanese experience (Tsutsumi 1999) and the record of the hybrid Dutch covenants (see below) illustrate (and it may be no coincidence that the cultures of these countries are particularly conducive to voluntary approaches). But initial reports of a number of other 'successes' have been heavily criticised.

Harrison (2001), for example, argues that first reports of environmental benefits of many voluntary programmes relative to reference years almost certainly exaggerate programme effectiveness: 'Since participation is voluntary, claims of benefit beyond "business as usual" can be viewed with less confidence since firms may be selectively signing on only to do what they would have done anyway.' Another general survey has also concluded that many of the first generation of negotiated agreements have not been markedly successful (Gibson 2000). Even the claim that negotiated agreements tend to reduce administrative burdens is not confirmed by existing empirical evidence or analysis (OECD 1999b: 131).

The one example of negotiated agreements, namely the Dutch environmental covenants, that have been demonstrably successful defy formal classification. They represent (in European terms) an unusual hybrid, since they address both collective and sector-wide environmental issues and are legally binding on individual enterprises through the permit system, and are thus intimately linked to mainstream command-and-control. Rather than playing an ancillary or supporting role (as is the case with many European collective agreements), they are a key component of Dutch environmental policy. Under those agreements enterprises draft environmental plans for each of their plants (identifying environmental targets and how they propose to achieve them), which are set at a level intended to achieve the framework targets established in the National Environmental Plan set by the national government. Local permit authorities have a role in approving and reviewing the individual environmental plans, and failure to establish or meet plans would result in tougher permit requirements being imposed. Thus companies can be made individually legally liable.[14]

A critical distinction between the Dutch approach and almost all other negotiated agreements is that in the case of the former the government sets non-negotiable goals based on collective performance objectives previously established under the National Environmental Plan. As a result 'government and industry have compatible objectives for negotiations: to find the most cost effective means to achieve those goals . . . however, when the goals themselves are negotiable, strategic behavior and "capture" by industry are serious threats' (Harrison 2001: 222). While only limited empirical study of the Dutch approach has taken place, work by the European Environment Agency suggests that the Dutch covenant with the chemical industry did achieve substantially better outcomes than a projected business-as-usual trend, and that it was environmentally effective with regard to at least 33 of the 61 chemicals studied (EEA 1997).

---

14   This description is taken from Borkey and Lévêque 1998.

Summary

There may be a variety of reasons for the modest success of many of the first generation of voluntary approaches (unilateral, public voluntary and negotiated agreements) more generally, including: the central role of industry in the target-setting process, the scope for free-riding, the uncertainty over regulatory threats, non-enforceable commitments, poor monitoring and lack of transparency. In turn, the manifest deficiencies in the design of first-generation instruments suggest a number of lessons about how to design such approaches better in the future. For example, the OECD has identified a number of 'success' criteria (see Box 5), which, if followed, may achieve more positive results.

Beyond the general list given in Box 5, a number of particular issues emerge from the various empirical studies and analyses conducted to date.[15] First, there are well-known and readily identifiable benefits in including third parties in the process of developing and overseeing such approaches:

> an open, transparent negotiating process can reduce the possibility of regula-
> tory capture and lessen the harmful effects of excluding third parties. When
> negotiating positions are open and known to the general public, opportunities
> for capture decrease significantly. Agency officials are forced to be more
> accountable to the public and to other third party interest groups. Further-
> more the information that third parties bring to the negotiations may inform
> the decision making process, expand the scope of the debate, and lead to a
> better outcome (Krarup 2001).

On the other hand, the exclusion of third parties and the lack of an open, transparent process may speed up the regulatory process significantly. It is these features of many of the early agreements that serve to lower transactions costs and provide the greater flexibility and the avoidance of conflict between the industry and the regulator, which industry seeks. Again, the more onerous the monitoring and enforcement provisions, the higher transaction costs will be. As Higley *et al.* (2001: 11) point out:

> voluntary approaches can be flexible, and therefore a cost effective way of
> developing environmental policy, but this advantage comes at the expense of
> public participation and ultimately, environmental effectiveness. Conversely,
> if invested with the proper procedural safeguards, voluntary approaches can
> be an effective policy instrument for achieving ambitious environmental goals,
> but this effectiveness comes at the expense of flexibility.

A second broad point is that voluntary approaches are often best used when an environmental problem is in its early stages and it is premature to regulate it directly. As the OECD (1999: 134) has pointed out:

> in this regard voluntary approaches can be regarded as a policy instrument
> with a transitional function, i.e. to work until time is ripe for other regulations
> to come into force. They are particularly suitable for this role, since they are
> likely to generate soft effects and learning, and hence can help improve the
> future design of more traditional instruments.

Third, as we have argued elsewhere (Gunningham and Grabosky 1998: Ch. 6), the weaknesses of voluntarism can often be compensated for, and its strengths enhanced, by

---

15   Of these the most influential include OECD 1999 and Covery and Lévêque 2001.

Clearly defined targets. The targets should be transparent and clearly defined. Moreover, the setting of interim objectives is crucial since they permit all the parties to identify difficulties arising at an early stage of implementation. Government should take the initiative in target-setting, obtaining information from a variety of sources, such as benchmarking.

Characterisation of a business-as-usual scenario. Before setting the targets, estimates of a business-as-usual trend—what the emission levels or other target variables are likely to be, given natural technical progress within the industry in question—should be established in order to provide a baseline scenario.

Credible regulatory threats. Made at the negotiation stage, a threat of regulation by public authorities provides companies with incentives to go beyond the business-as-usual trend: 'when the agency has a stick waiting in the wings, industry is more likely to accept the carrot. That is, the existence of a regulatory threat as an alternative to an agreement, promotes voluntary action by industry' (Segerson and Miceli 1998).

Credible and reliable monitoring. Provisions for monitoring and reporting are essential for keeping track of performance improvements. They constitute the key for avoiding failure to reach targets. Monitoring should be made at both the company level and the sector level in the case of collective voluntary approaches. In certain contexts, monitoring by independent organisations may be used.

Third-party participation. Involving third parties in the process of setting the voluntary approaches objectives and in its performance monitoring increases the credibility of voluntary approaches. More generally, environmental performance should be made public and transparent. It provides industry with additional incentives to respect their commitments.

Penalties for non-compliance. Sanctions for non-complying firms should be set. This can be achieved by making either binding commitments or linkages between voluntary approaches commitments and regulatory requirements (e.g. the integration of voluntary approaches requirements into operating permits).

Information-oriented provisions. In order to maximise the informational soft effects of voluntary approaches, support for technical assistance activities, technical workshops, the publication of best-practice guides, etc., should be promoted.

Provisions reducing the risk of competition distortions. In the case of collective voluntary approaches, safeguards against adverse effects on competition could be provided by notification of new voluntary approaches to anti-trust authorities.

Box 5   *Recommendations on the design of voluntary approaches*

Source: OECD 1999: 134-35; Covery and Lévêque 2001

combining it with most, but not all, forms of command-and-control regulation.[16] Such an underpinning of government regulation may be particularly important in curbing free-riding.[17] Fourth, voluntary approaches do seem to generate major positive soft effects such as collective learning, generation and diffusion of information, learning by doing and demonstration effects, and consensus building. Since many VAs aim at increasing environmental awareness of the industry rather than short-term environmental impacts, these learning and innovation effects should not be dismissed lightly.

Finally, the evaluation of negotiated agreements in particular requires a dynamic analysis: the second generation of such agreements are somewhat different from the first, and considerably more likely to provide public interest benefits. Much more specific targets now tend to be set by government rather than vaguer goals being determined by industry, government negotiators are much more sensitive to the risks of setting targets that merely reflect improvements that would happen anyway, and there is a movement towards linking negotiated agreements with other policy instruments, such as taxes, or to complement rather than replace existing regulations. Greater efforts are also being made in terms of transparency and third-party input. Whether these developments will justify the faith of advocates of voluntary approaches, and whether the additional transaction costs of building in essential checks and balances will render such instruments too costly, remains to be seen. But that question must be asked in the context of a comparison with the other available alternatives. In the light of the relatively modest success of most traditional regulation at least, the yardstick for comparison may not be that high.

Beyond all else, voluntarism's trump card is its political acceptability. It may raise the hackles of environmental groups for good reason, since it not only curtails the role of both government and other non-government actors but also places considerable faith in the business community, which, it has been pointed out, is 'the last place many would look for altruism' (Harrison 2001: 237). By the same token it is also far more palatable to economically powerful industry groups than almost any other option, and it is relatively

16 Sometimes this will include direct regulation, sometimes the threat of regulation that will be triggered automatically if certain voluntary targets are not met (the French and German agreements on packaging waste recycling are backed by decrees, specifying the regulations that would come into force in case of voluntary agreement failure) and sometimes by combining VAs with *some* economic instruments (Gunningham and Sinclair 1999a). For example, the Danish Agreement on Industrial Energy Efficiency involved a policy mix combining VAs on reductions of carbon dioxide and sulphur dioxide with taxes and subsidies. The most important incentive provided was a substantial carbon dioxide tax rebate for those companies which entered a negotiated agreement. Such agreements have been reached both with individual companies and with groups of companies within a sub-sector. Companies entering the agreement commit to undertake an energy audit, usually conducted by a certified consultant, together with an independent verification. An action programme is developed on the basis of the audit report. All energy-efficiency projects with a payback period of less than four years must generally be carried out as part of the plan (although alternative projects can be proposed). In the event that companies fail to meet their obligations in the agreement, the Danish Energy Agency can cancel the agreement and the tax rebate will be annulled (Krarup and Ramesohl 2000).

17 Conversely, 'regulatory components provide voluntary approaches with safeguards against their main shortcomings: namely, low expectations in their environmental targets, weak enforcement provisions, and the lack of credible and efficient monitoring and reporting requirements' (OECD 1999: 11).

cheap to implement (particularly if third parties are cut out of the process). For these reasons the temptation for politicians to escape from uncomfortable conflicts through greater use of voluntarism, irrespective of its merits or demerits, often proves over-whelming.

Regulatory flexibility

A substantial number of large enterprises now recognise their obligations to comply with environmental regulation and do so irrespective of the likelihood of detection or sanction (Fischer and Schot 1993; Gunningham et al. forthcoming), in contrast to an earlier generation which frequently sought to evade it. Increasingly, such enterprises are also developing environmental strategies that incorporate pollution prevention, internal compliance auditing and compliance assurance programmes (Coglianese and Nash 2001). Some are also actively seeking win–win outcomes which combine environmental and economic goals and which are capable of taking them well beyond compliance with existing regulation (Hart 1997; Reinhardt 2000).

These developments raise considerable challenges for the traditional system of regula-tion. That system was concerned primarily with bringing enterprises up to a minimum legal standard, a function that is still important with regard to laggard enterprises, but which is increasingly irrelevant or counterproductive in relation to companies that are ready, willing and able to go beyond compliance. For these enterprises, the challenge is to design environmental policies that reward, facilitate and encourage them to do so. The broader challenge is to shift the entire regulatory compliance curve (embracing laggards at one end and leaders at the other) in the direction of improved environmental perfor-mance.[18]

Regulatory flexibility is one term applied to a range of policy initiatives that seek to address these challenges by providing incentives for individual enterprises to achieve beyond-compliance outcomes building on an existing regulatory framework. In this respect it may be distinguished from other voluntary approaches described above: namely, unilateral approaches (which have no government involvement), public volun-tary programmes (in which government provides non-regulatory incentives, and the improvements are independent of and additional to existing regulatory obligations), and negotiated agreements (which apply to entire industry sectors, and are an alternative to existing or planned regulation). The key feature, then, of regulatory flexibility is the extent to which it integrates with regulation.

Early efforts at regulatory flexibility focused on a shift away from specification- or technology-based environmental regulations[19] towards performance standards (Davies

18   This is increasingly recognised by the US EPA, which now classifies businesses as leaders (going well beyond compliance), mainstreamers (achieving compliance but little more) and 'those who have been left behind' (who do not achieve even the legal minima). See US EPA 1999.

19   Technology-based standards have most commonly been adopted under pollution law in some jurisdictions, including the US. Technology-based conditions are usually a variant on a require-ment that 'best available technology' or 'commonly available technology' must be used to curb pollution or environmental harm, sometimes with the qualification that only economically

and Mazurek 1996).[20] The logic was that performance standards can foster innovation because they leave an enterprise free to respond to a regulator's requirement in the way it thinks best (Porter and van der Linde 1995: 129).[21]

However, performance standards still have a substantial limitation: namely, they only require enterprises to achieve minimum standards and provide no incentives or encouragement to go beyond those minima. They do not encourage continuous improvement or industry best practice. Nor do they directly encourage enterprises to develop an environmental culture or to 'build in' environmental considerations at every stage of the production process. However, as indicated above, there are many other enterprises that, potentially at least, could achieve far more than those minima. For these enterprises, process and management system standards may provide not only considerable flexibility and enable enterprises to devise their own least-cost solutions but can give them direct incentives to go beyond compliance. We have engaged in an extended treatment of this issue elsewhere (Gunningham and Johnstone 1999: Chs. 4, 5), and for present purposes confine ourselves to two issues of direct concern to the principle and practice of regulatory flexibility: the potential contribution of systems-based regulation; and the design of regulatory flexibility initiatives.

Systems-based regulation

As indicated in Chapter 2, an EMS is a management tool intended to assist the organisation in achieving environmental and economic goals by focusing on systemic problems rather than individual deficiencies. The basic elements of such a system include the creation of an environmental policy, setting objectives and targets, implementing a programme to achieve these objectives, monitoring and measuring its effectiveness, correcting problems, and reviewing the system to improve it and the overall environmental performance. In future, the most popular form of EMS will almost certainly be one that complies with the EMS standard, ISO 14001. This standard, introduced in response to the demand for a single, 'off-the-peg' internationally recognised management standard, is likely to prove one of the most significant developments in the field of environmental management (and possibly of regulation) in many years (ISO 1996; Tibor and Feldman 1996).

achievable technology must be used. In theory, best available technology may be used to determine the appropriate level of an emissions standard, but in practice it often becomes a de facto prescription for specific technological solutions.

20 A performance standard is one that specifies the outcome of the environmental improvement but which leaves the concrete measures to achieve this open for the employer to adapt to varying local circumstances. That is, rather than specifying exactly how to achieve compliance, a performance standard sets a general goal and lets each enterprise decide how to meet it.

21 In evidence Porter and van der Linde cite the relative success of the Scandinavian and US governments in achieving emissions reductions in the pulp and paper industry. The Scandinavian companies, under a performance-based regime, 'developed innovative pulping and bleaching technologies that not only met emission requirements but also lowered operating costs'. US companies, in contrast, did not respond to regulation by innovating because US specification- or technology-based regulations did not permit companies to 'discover how to solve their own problems' (Porter and van der Linde 1995: 129).

A particular attraction of EMSs is their capacity to move corporate thinking on environmental performance from the sort of compartmentalisation that characterised the earlier generation of pollution control instruments (vertical standards addressing discrete areas of activity) to a horizontal standard that cuts across the functions of the organisation, and integrates environmental considerations with other corporate functions and imperatives. If this occurs, then 'cost, efficiency, productivity and environmental performance all become part of the same decision-making process' (Knight 1994: 45). On the other hand, there remains some ground for caution: lasting cultural change is difficult to achieve within an organisation, and, even if it is achieved, a causative link to tangible environmental improvements are not necessarily clear (Coglianese and Nash 2001).

There is tentative, but at this stage not conclusive, evidence that an EMS approach has the capacity to bring about considerable improvements in performance; that 'by building internal awareness of environmental issues, firms can attain very substantial improvements in environmental performance' (Wells and Galbraith 1998: 35). However, the preliminary research does not suggest that EMSs will be uniformly beneficial, and indeed some enterprises (which see environmental practices as marginal to their strategic and competitive objectives) appear to treat EMSs as tools for external image manipulation rather than for genuine environmental improvement (Coglianese and Nash 2001). Indeed, it may even be companies that are struggling with environmental performance that are attracted to schemes such as ISO 14001 and the European EMAS (ENDS 2000d).

A number of regulatory flexibility initiatives have been introduced in a variety of jurisdictions that offer incentives for adopting EMSs. These include: fast-tracking of licences or permits, reduced fees, technical assistance, public recognition, penalty discounts under certain conditions, reduced burdens from routine inspections and greater flexibility in means permitted to achieve compliance. In return, industry is expected to go beyond compliance in environmental outcomes (and in some instances to engage in stakeholder dialogue, to be more transparent, and to take greater responsibility for the environmental behaviour of others in the supply chain).

A central *quid pro quo* for such regulatory rewards under most regulatory flexibility schemes is the agreement to have the EMS externally verified (i.e. by a third-party audit).[22] However, a danger of relying on ISO 14001 alone as an alternative compliance mechanism is that, beyond a requirement to meet existing legal obligations, nothing in ISO 14001 either measures or ensures tangible improvements. Consequently, enterprises with widely differing levels of environmental performance, even within the same industry sector, may all establish an EMS that complies with ISO 14001. Accordingly, any regulatory redesign built around an EMS should be used to complement rather than replace other regulatory tools.

22   In the USA in particular, the use of ISO 14001 as a regulatory tool has caused considerable concern to companies which fear, with some justification, that external third-party verification of the environmental management system could result in unfavourable information being made available to the public and regulators and that the latter might use this as a basis for prosecution. In particular, under US federal law firms cannot claim a self-evaluative privilege for the audits of their EMSs. However, recognising the disincentive to good environmental practice that this represents, the US EPA has substantially mitigated the problem with its 1995 policy statement: 'Incentives for Self-Policing: Discovery, Disclosure, Correction and Prevention of Violations', in which the EPA promises not to seek gravity-based, punitive civil damages against any company that discovers the violation through routine due diligence, if a number of conditions are met.

Designing regulatory flexibility

Many policy analysts argue that the new regulatory flexibility initiatives must be based on 'ISO Plus' rather than mere conformity with ISO 14001 (*Environmental Forum* 1999; Center for Strategic and International Studies 1998). Unfortunately, this has not necessarily been the case in practice.[23] Our own past analysis suggests that there are four key components necessary for the successful implementation of such regulatory flexibility initiatives (Gunningham and Grabosky 1998: 247). These are:

- That enterprises engaging in regulatory flexibility should adopt practices and processes that lead to the pursuit of beyond-compliance goals and include outcome-based requirements, the achievement of which can be measured through specific performance indicators[24]

- That there should be independent verification both of the functioning of their management system and of environmental performance under it (e.g. by a third-party environmental auditor), with the results or a summary of the results available to both the regulator and third parties such as community groups (transparency)

- That there should be an ongoing dialogue with local communities concerning beyond-compliance goals and the means of achieving them (this ensures the credibility and legitimacy of the process and enables third-party input and oversight)

- That there should be an underpinning of government intervention, acting as a safety net that comes into effect only when triggered by the failure of the other, less intrusive, mechanisms described above

Regulatory flexibility initiatives built around EMS are most fully developed in the United States under the Clinton–Gore administration's Reinventing Environmental Regulation initiatives (Clinton and Gore Jr 1995) and comparable initiatives. Project XL (for eXcellence and Leadership) is the flagship Reinvention initiative, whereby an applicant commits itself to exceed current emissions regulations in return for the EPA altering regulatory requirements to reduce the company's compliance costs. For example the EPA may approve a single comprehensive permit instead of requiring multiple permits that the applicant would otherwise have to submit to obtain regulatory permission to pollute air or water. The goal is to reduce regulatory costs and pollution by adopting 'customised' protection strategies for individual facilities. The primary criterion for acceptance into the programme is that firms must be able to demonstrate environmental results which are 'better' than those they could achieve under full compliance with present laws and regulations, in terms of eight specified criteria.

23 See also Crow 2000, noting at page 24 that 'less than half of [environment leadership programmes] require any proven pollution decrease (or other environmental outcome) either for entry or for staying in the program'.

24 Among the most important features of such systems are: incorporating pollution and waste prevention into core business practices; accounting for the total environmental impact of choices throughout the life-cycle of products and services; improving efficiency; considering environmental costs to society in business decisions; employing planning processes to illuminate pollution prevention and product stewardship opportunities; and continuously striving to improve.

Building on the experience of Project XL and of a number of other federal and state pilot projects, regulators are now moving towards a more systematic approach, designed to provide rewards and incentives for improved compliance and high environmental performance through a two-track system of regulation. Under this approach, enterprises (or at least enterprises with certain environmental credentials)[25] are offered a choice between a continuation of traditional forms of regulation on the one hand, and a more flexible approach (the central pillars of which are usually the adoption of an environmental management system, periodic internal environmental audits, and community participation) on the other. Examples of this approach include the Wisconsin Green Tier Proposal (Wisconsin Department of Natural Resources 2000) and the US EPA's Performance Track (US EPA 2000a) (see also Crow 2000).

The ultimate test of the success or otherwise of regulatory flexibility initiatives such as the above is an empirical one. The environmental policy literature abounds with seemingly attractive reform initiatives that failed to perform as predicted. And the early evidence is not all positive. Most of the new initiatives seem to 'operate more or less at the margins of the existing regulatory system, tinkering with incremental changes' (Crow 2000: 29). Project XL, in particular, has met with sustained criticism from a variety of quarters as serious problems mount up. The programme has attracted far fewer applications than anticipated, applicants have proposed fewer innovations than EPA had hoped, the approval process has taken longer than expected, and the administrative law implications of what is essentially 'contract-based regulation' have not been worked through.[26] Overly high transactions costs, a failure to overcome mutual mistrust, the lack of a statutory base, and a vague definition of 'better' results have also been identified as particular weaknesses.[27]

Perhaps Project XL's most fundamental weakness is that it is not well integrated into the legal system. Indeed, it does not have any legal basis. XL agreements involve government agreeing to waive strict application of some environmental laws in return for higher overall environmental performance (not required by environmental law). But lacking legal authority to do this, such agreements can be challenged successfully by environment groups and other NGOs. Given their mistrust of deals between government and industry, the likelihood of such suits being brought in relation to an XL agreement is high. The result is that business is very reticent to enter into such agreements. because it does not have confidence that government can honour its side of the bargain. In the culture of the US, aptly described as one of 'adversarial legalism' (Kagan 2001; Kagan and Axelrad 2000), Project XL, with its lack of any formal legal basis and its vulnerability to third-party citizen suits, has struggled to make any positive impact.

But it would be dangerous to generalise from the particular limitations of one early and, arguably, poorly designed pilot programme. Despite their very considerable poten-

---

25   A demonstrated record of above-average (or 'good') environmental performance and no major violations is a vital 'gateway' requirement which serves the important purpose of denying access to track-two regulation to firms that, given their past environmental record, would be most likely to abuse it.

26   The best account of the administrative law dimension of this evolving saga is Freeman 2000. See also Seidenfeld 2000.

27   For general critiques of Project XL, see Steinzor 1996; Susskind et al. 1998. For a broader critique of environmental flexibility initiatives, see Davies and Mazurek 1996. For a detailed and highly insightful analysis of how one XL experiment failed, see Marcus 1999.

tial, the jury is still out on the strengths, weaknesses and, ultimately, the success of EMS-based regulatory flexibility initiatives more generally. It will be some time before we know whether and, if so, to what extent, the benefits of the various initiatives outweigh the costs and whether they will in fact overcome many of the problems of traditional forms of regulation.[28] It will also be some time before it emerges whether the sceptics are correct in questioning why so many resources are being devoted to making the top 20% (or perhaps only the top 5%) even better, rather than concentrating on the most serious problems or on under-performers.

In Chapter 8 we make our own contribution to the limited empirical literature through our case study of regulatory flexibility in the Australian state of Victoria. This involves the use of two complementary instruments: accredited licensing and environmental improvement plans. The former is intended for companies committed to 'best-practice' environmental management, and, as with the American initiatives, the *quid pro quo* for providing regulatory flexibility and relief is the development and implementation of a number of mechanisms, the most fundamental of which is an EMS. In some respects, the environmental improvement plan is even more ambitious, seeking to engage not only leaders but also laggards in 'facilitative regulation', which involves a process-based approach in conjunction with community participation and regulatory oversight. As we will see, some aspects of this latter initiative have been strikingly successful, which regrettably cannot be said for the former. From this experience, we seek to draw some broader lessons for the future design and role of regulatory flexibility initiatives internationally.

Economic instruments

According to environmental economics, environmental degradation results from a failure to fully value environmental endowments in the market. In this paradigm, it is because no price emerges to reflect their scarcity value that 'the market fails to fulfil its most fundamental of functions, that of rationing scarce assets efficiently' (Higley *et al.* 2001). The conventional solutions to this problem are either to introduce market signals by charging appropriately for the use of such scarce resources (e.g. through a tax or charge) or by determining the amount of pollution that the environment can assimilate, allocating rights to those emissions, and letting the market determine the appropriate price.

Economists argue that economic instruments are usually more cost-effective than direct regulation, in large part because they give companies more flexibility as to how they achieve resource productivity and prevent pollution. Moreover, as the World Busi-

---

28 For a summary of empirical evidence, see the National Database on Environmental Management Systems, which includes more than 50 pilot facilities that are implementing EMS, at www.eli.org/isopilots. The Multi-State Working Group on the Environment (MSWG) (a collection of US states that have formed a nationwide partnership) is currently gathering credible uniform information to determine whether implementation of EMS actually enables pilot companies to better meet their legal obligations and achieve broader environmental performance goals.

ness Council for Sustainable Development asserts, 'they provide continuous incentives to producers to conserve resources, prevent pollution and step up technological and organisational innovation [and] are the most direct way of changing producer and consumer behaviour toward more efficient resource use' (WBCSD 1995). For all these reasons, economic instruments are a potentially important policy tool.

It should be noted that some, but far from all, such instruments are more suited to dealing with large enterprises than with SMEs. This is particularly true of tradable pollution permits, where transaction costs and the size of basic units that it is practicable to trade preclude their application to any except substantial point-source polluters.[29] Taxes and charges, on the other hand, have greater applicability to SMEs, not least because costs can be passed down the supply chain relatively effectively, reducing the possibilities of evasion. And subsidies are almost ideally suited to SMEs, particularly those that are economically marginal and far less likely to respond to negative incentives.

The full range of economic instruments that might be deployed in environmental policy is quite extensive, and ranges from the removal of perverse incentives through the use of taxes, charges and market creation, to the use of liability rules,[30] and includes both property rights[31] and price-based instruments. It is beyond the scope of this project to give an extended treatment of the use of such economic instruments, and as non-economists we would be loath to do so.[32] However, it is important to explain why, notwithstanding the impressive claims made for what such instruments can achieve in theory, they have only been invoked to a very limited extent in practice, and with variable results.

In the remainder of this section we focus on the use of price signals in the shape of taxes or charges, property rights in the form of tradable permits, and supply-side instruments in terms of subsidies, as representative examples of economic instruments that can be applied in the area of point-source pollution control. In doing so, we highlight a number of themes relating to their use and misuse and their likely practical policy contribution.

A pollution tax or charge should ideally be set so as to assign to polluters an appropriate price for the pollution they emit. Aware of this price, the polluter has an incentive and an opportunity to design the least-cost method to reduce emissions up to the point where it is more rational to pay for the pollution. In formal terms, a tax measure has the static efficiency advantage that it can achieve a target level of pollution abatement at least cost by equalising marginal abatement cost among polluters.[33] The greatest advantage that environmental taxes and discharge fees have over conventional regulation is that both instruments allow firms to make individual choices about environmental performance free from outside interference and in doing so give them an incentive to innovate and reduce pollution at the lowest cost.

29 Other instruments, such as environmental taxes, can be applied more broadly, although even here the 'rational economic actor' model that underpins such mechanisms more closely approximates the behaviour of large corporations than that of many small businesses.

30 For a comprehensive description, see Panayotou 1998. For a succinct but excellent current analysis, see Pearson 2000.

31 On the extent and means by which property rights can form a crucial element of optimal environmental policy, see Rose 1997.

32 The value of such instruments in the environmental context has already been examined in a number of OECD publications. See in particular OECD 1998.

33 The classic treatment is Baumol and Oates 1998.

However, notwithstanding their considerable theoretical attraction, such instruments have only been used to a very limited extent. Despite their criticisms of command-and-control, regulated businesses have consistently opposed the introduction of taxes and charges, preferring the certainty of regulation to the uncertainty of novel approaches. More fundamentally, they have worried that pollution taxes would be used by politicians as an underhand way of raising the overall tax burden on them, rather than as a genuine environmental tool.[34] To accept even a modest tax, they feared, would be the 'thin end of the wedge' because over time politicians would succumb to the temptation to ratchet such taxes upwards, even where this could not be justified in environmental policy terms.

In practice, such tools have often been misused to produce the worst of both worlds. On the one hand, politicians have been reluctant to impose the full amount of tax necessary to implement the polluter-pays principle for fear of a political backlash. On the other hand, the taxes are often imposed more with an eye on boosting the government's overall revenue than on providing a more efficient and effective replacement to command-and-control. This is certainly the experience in Europe, where there has been a greater reliance on effluent fees and charges than elsewhere. Many of these instruments are demonstrably designed for financial purposes (revenue generation) and supplement rather than replace direct controls, with the charges being set well below the levels necessary to provide the appropriate incentives for pollution abatement (Pearson 2000: 164).

Other design faults may also become apparent in relation to such taxes. In the UK, for example, a landfill tax credit scheme has been applauded by economists as having considerable static and dynamic efficiencies, and there are moves to introduce comparable taxes in relation to pesticide use and the use of virgin aggregates. However, the most recent and detailed evaluation of the landfill tax is that it has fundamental weaknesses: it is dominated by the waste industry; is prone to conflicts of interest between waste companies and local councils; and had been defrauded of substantial sums (*Guardian* 2001).

Four broader objections can be made to the use of taxes and charges.[35] First, there is considerable difficulty in setting a tax or charge at the right level. This is because the costs and choices facing polluters may not be known to policy-makers (and, in any case, may be subject to lobbying to minimise business costs). This problem might be overcome by trial and error (which may disrupt investment plans) or by establishing a progressive scale of charges, to be increased over time. Second, where prices are relatively inelastic due to limited input substitutability, costs may simply be transferred to final consumers with no consequential environmental benefit. Alternatively, the size of the tax or charge would need to be very large and thus undermine the cost-effectiveness of the instrument.[36] Third, firms may not respond rationally to price signals. Where taxes or charges represent only a small proportion of outlays, costs might simply be ignored or not noticed. And, fourth, taxes and charges may be perceived as legitimating or condoning

---

34 This criticism could be overcome by requiring that the funds be returned for environmental improvements in the sector from which they have come, but, although that approach has been typical in France, it is not common elsewhere.

35 This paragraph is adapted from Gunningham and Grabosky 1998.

36 For example, the consumption of petrol for private motor vehicle use is very price-inelastic; because of the unavailability of alternative fuels, the price of petrol would need to be very high before consumers would substantially change their driving behaviour.

environmentally harmful behaviour. The notion that the state will allow pollution at a price strikes some members of the community as an inferior regulatory strategy to one that condemns environmentally deleterious activity outright.

Nevertheless, if carefully designed, taxes and charges can be both efficient and effective. For example, in Germany the government created sufficient tax incentives for the introduction of catalytic converters. They were careful to ensure that the programme was revenue-neutral in order to overcome scepticism that this strategy might be merely a tax-grab. The tax differentials were carefully calculated, consistent, well balanced and easy to administer. More important, ordinary people could understand them and act accordingly when they bought a car. For these reasons they were successful in inducing people to buy (relatively) environmentally friendly cars. This fiscal instrument worked in parallel with command-and-control emission standards: the tax incentive increased demand for environmentally friendly cars. This induced the car industry to produce such cars. Following this market reaction, government was able to set tighter standards for emission limit values without opposition (Schnutenhaus 1995). The introduction of a nitrogen oxide charge on energy production in Sweden some years ago, and the use of carbon dioxide taxes in Scandinavia, are other examples of the successful application of environmental taxes (Olivecrona 1995a, 1995b).

Tradable permit schemes, in contrast, adopt a property rights approach to achieving efficient pollution control. Under such schemes the regulatory authorities set a target as to how much pollution will be allowed in an industry or an area over some fixed price period. This quantity is then divided into 'rights' or permits, which are auctioned or otherwise distributed to polluters. Firms that keep their emissions below the allocated level are free to sell their excess. Enterprises that anticipate polluting beyond their allocated level must buy permits from other polluters who have reduced their emissions to below their allocation and have pollution credits to sell (or purchase them from periodic or one-off auctions). In either case the trading system provides incentives to participating firms to reduce emissions so far as they efficiently can, and also gives them the flexibility to do it in the least-cost manner. In essence, the government has created limited property rights, providing the foundation for a private market.

But, again, despite the theoretical elegance of such schemes, and notwithstanding their attractions in giving firms greater flexibility in tailoring responses to their individual circumstances, they confront a number of practical problems. In particular, they are *not* self-enforcing. Rather, government must be in a position to enforce the allowable emissions of individual permits to prevent abuses by free-riders, in a similar fashion to the enforcement provision of traditional regulatory instruments. There are particular difficulties in monitoring and enforcing permits when there are a large number of small, disparate polluters or there are mobile sources of pollution, such as vehicles, or there are non-point sources of pollution, such as methane emissions from farms.

Perhaps unsurprisingly given the practical obstacles, only a modest number of market creation schemes have so far been introduced. Some, such as the US acid rain permit trading programme (see Box 6), have the capability to work with considerable effectiveness, albeit with far less equity (US EPA 1996). Indeed, this programme is widely regarded as a great success (Burtraw 1998; Burtraw et al. 1997; Joskow et al. 1996), and is the proposed model for an international permit trading system in relation to carbon emissions and global warming.

Background and compliance problem

Previous legislative attempts to reduce sulphur dioxide emissions causing acid rain have focused on imposing new point-source-specific emissions rate limitations: for example ,on emissions from newly constructed electricity-generating facilities (powered by fossil fuel). The effluent limitation, by mandating an effluent rate reduction, effectively dictated technological choices in a typical command-and-control fashion. This meant that compliance through the use of process changes or demand reduction was precluded. Operators of new facilities could not switch to cleaner fuels to achieve emissions reductions, or use pollution prevention technologies. Furthermore, the limitations were only applicable to new facilities, while existing (more polluting) facilities were not subject to the standards.

The compliance innovation

The acid rain programme established by the Clean Air Act Amendments of 1990 instigated two innovative features that represent dramatic departures from conventional environmental regulation. First, an overall industry-wide cap on emissions of sulphur dioxide has been set at approximately 'one half of historic levels'. Secondly, the programme allows operators of affected facilities (including existing facilities) to trade emissions allowances between their own facilities or with other utilities in order to save costs in achieving the national emissions cap. Thus, under the new programme a facility has the freedom to implement abatement measures that depart from an engineering prescription of the cleanest possible technology for that facility (such as the use of low-sulphur coal fuel blending). In order to do this, it must find alternative ways of reducing emissions or it must compensate another facility to reduce emissions accordingly.

The results

The results would suggest that the more general an overall performance standards approach can be (as opposed to a rate-based approach), the more technologies can be utilised in compliance, and the higher the likelihood of breakthroughs and cost-effectiveness. Facts illustrating this include:

- In 1995 and 1996 utilities over-complied by emitting approximately 30% less sulphur dioxide than the programme's emission cap stipulated.

- A US geological survey study found a decrease in precipitation acidity of 10%–25% in the north-east in 1995.

- An EPA study estimates the value of the human health benefits of the programme at between US$3 billion and US$40 billion nationwide.

- The estimated compliance costs of US$2 billion or less have halved the US$4 billion per year spent under the previous regulatory system, in just two years of full implementation.

- There was virtually 100% compliance in the first two years without the need for enforcement.

- There has been a reduction in litigation.

- The programme requires very few regulatory staff to manage.

- The price of an emission allowance of one ton of SO_2 has dropped from US$750 to around US$100.

- The technological flexibility has allowed facilities to achieve compliance with the emissions cap without being overly dependent on inter-utility trading of emission allowances.

Box 6   *The acid rain programme*

Source: Swift 1997: 17-25

Notwithstanding the success of the United States acid rain trading programme, other schemes have suffered serious design faults,[37] with the result that very few trades have actually taken place and monopolistic and anti-competitive behaviour has emerged.[38] There is a dilemma in increasing the complexity of the design and operation of permit schemes to address unintended consequences, in that they have the potential, through greater administrative, compliance and enforcement burdens, to undermine theoretical efficiencies. It is difficult to escape the conclusion that, despite potential efficiency gains, market creation may be restricted to applications where the use of permits can easily be monitored and verified, and where there are good trading prospects. In these circumstances, well-designed schemes have the capacity to deliver substantially reduced pollution loads and a substantially lower cost to industry.[39]

Finally, we turn to supply-side incentives. These refer to direct or indirect payments, including tax concessions and subsidies, conditional on desired conduct. For example, they include: tax concessions for the purchase of cleaner-production technology; tax deductions for the expenses of environmentally responsible activity, such as mine site remediation; and lower tax rates on preferred products or materials, such as energy-efficient cars or unleaded petrol. The use of such subsidies is usually deplored by environmental economists because (by blocking the internalisation of external cost in prices) they contravene the polluter-pays principle and are for this reason inefficient, and because they are a drain on public revenue. Treasury departments often oppose such incentives because of the difficulties in classifying products and practices that might attract the subsidy, and because they have the potential to lock in current technologies at the expense of as yet unknown alternative solutions.

Nevertheless, some economists have suggested they may have merit in two situations. First, because of imperfect information, deficiencies in capital markets, or the public-good characteristics of technology, 'there may be insufficient investment in the production and dissemination of environmentally friendly technology, technique and products. If such market failures are pervasive, then an argument can be made for government subsidies for the development and use of such technologies and products' (Pearson 2000: 162). In these circumstances, the international experience suggests that supply-side incentives can be an effective spur to cleaner-production investment. For example, the OECD Technology and Environment Programme found that 'natural technological

---

37 For example, both the EU carbon–energy tax and the trading of sulphur quotas as a means for implementing the EU Large Combustion Plants Directive have been costly failures. See Howes *et al.* 1997: Ch. 9.

38 With many permit schemes in operation, the actual number of trades taking place between firms is well below expectations; this has perplexed policy-makers. One market for effluent permits experienced only one trade in six years (A. Moran, 'Tools of Environmental Policy: Market Instruments versus Command and Control', in Eckersley 1995). Explanations include: (1) insufficient players for a competitive market to emerge; (2) too many regulatory controls which inhibit trading by making it costly; (3) abatement technologies are limited and require discrete amounts of investment; this results in 'lumpy' investments which would need to be matched by an equivalent parcel of permits; (4) the initial allocation of permits may be overly generous, resulting in firms having more than enough permits for their needs and therefore no incentive to trade; and (5) hoarding of permits to limit new entrants or drive out competitors.

39 See in particular the experience of the South Coast Air Quality Management District, Southern California RECLAIM programme evaluated in Leyden 1997.

evolution occurring in industry has not forced environmentally-protective technologies to be designed or employed. Governments will need to promote cleaner production and products' (OECD 1995).

A second circumstance where subsidies may be economically defensible is where they are linked to achieving a pollution control objective. Here

> the authorities establish a base level of pollution for a firm and then offer the firm a subsidy of a fixed amount per unit of abatement below that level. The subsidy is the carrot . . . From a firm's perspective, continuing to pollute at its base level now carries an opportunity cost—the subsidy foregone. A profit-maximizing firm will then decide to undertake pollution abatement from its base level up to the point where its marginal abatement cost equals the subsidy rate, and the subsidy accomplishes the same abatement for the firm that an equivalent tax would have (Pearson 2000: 161).

Even where subsidies may be justifiable, they may not be effective. Research data indicates a strong correlation between the level of supply-side incentives and their uptake. Nevertheless, in recent years such approaches have become increasingly sophisticated. For example, the Netherlands has an accelerated depreciation programme for specific clean technologies, and Costa Rica now runs a transferable reforestation tax credit scheme under which land-holders receive a tax credit for keeping their land forested or for returning land to native species cover.

To summarise, there is evidence that economic instruments, carefully designed, can make a valuable contribution to specific environmental problems, and in some circumstances (as with the acid rain programme) have proved far more successful than previous policies. However, the extreme claim by some economists that command-and-control regulation is inevitably inefficient, or less efficient than economic instruments such as effluent taxes and marketable pollution permits, has been challenged both as a matter of economic theory and of experience.[40] In particular, market-based solutions are not well suited for all institutional and technological contexts, particularly where monitoring costs are exorbitant. 'In that circumstance, command-and-control regulations may be both more efficient *and* more effective. To the extent they are replaced by market mechanisms, it should only be after careful, case-by-case examinations of expected costs and benefits, including implementation and monitoring costs' (Cole and Grossman 1999: 936). More generally, the answer to how and when policy-makers should prefer economic mechanisms to command-and-control will be context-specific (Cole and Grossman 1999).

---

40 In particular, it has been argued that the prevailing view of alternative regulatory regimes is oversimplified in at least three ways: (1) it overemphasises the difference between command-and-control regulations and economic instruments; (2) it conflates nominal and relative economic efficiency in comparing alternative regulatory regimes; and (3) it tends to be ahistorical and acontextual, ignoring changes over time in marginal costs and technological capabilities. Regulatory institutions especially tend to assume 'perfect (and incidentally, costless) monitoring or they assume that monitoring costs are the same regardless of the control regime that is chosen' (Cole and Grossman 1999: 936).

Informational regulation

An increasingly important alternative or complement to conventional regulation is what is becoming known as 'informational regulation' (Sabel *et al.* 2000), which has been defined as 'regulation which provides to affected stakeholders information on the operations of regulated entities, usually with the expectation that such stakeholders will then exert pressure on those entities to comply with regulations in a manner which serves the interests of stakeholders' (Kleindorfer and Orts 1996). In contrast to command-and-control, informational regulation involves the state encouraging (as in corporate environmental reporting) or requiring (as with community right-to-know [CRTK]) the provision of information about environmental impacts but *without* directly requiring a change in those practices. Rather, this approach relies on economic markets and public opinion as the mechanisms to bring about improved corporate environmental performance. As such, informational regulation 'reinforces and augments direct regulatory monitoring and enforcement through third-party monitoring and incentives' (Kleindorfer and Orts 1996: 1).

Informational regulation is targeted almost exclusively at large enterprises, and in particular at public companies (which are vulnerable to share price and investor perceptions) and those that are reputation-sensitive, because is it essentially these types of enterprise that are most capable of being rewarded or punished by consumers, investors, communities, financial institutions and insurers on the basis of their environmental performance. The overall strategy is to empower these groups to use their community and/or market power in the environmental interest by providing them with sufficient quality and quantity of information to enable them to evaluate a company's environmental performance. Such a strategy becomes even more effective as companies recognise the importance of protecting their 'social licence' and the need to improve their environmental performance in order to do so (see Chapter 7).

There have been a number of experiments with the use of informational regulation, which have demonstrated its potency even in circumstances where conventional command-and-control regulation is weak. The most striking of these is in Indonesia, where regulatory resources are grossly inadequate and there is no credible enforcement of environmental standards. Under the PROPER PROKASIH programme, priority polluters are required to negotiate (legally unenforceable) pollution control agreements with teams comprising public agencies, environment groups and regional development groups.[41] Regulators rank the performance of individual facilities using surveys, a pollution database of team reports, and independent audits. An enterprise's pollution ranking is readily understood by the public, as it is based on a colour coding (gold and green for the best performers, black, blue and red for those not in compliance). The programme has been very successful in improving the environmental performance of participating firms. A recent study, which examines the programme over time, suggests that community pressure and negative media attention, and increased likelihood of obtaining ISO 14000 certification, are the major stimuli for improved environmental performance. However, the provision of increased environmental information to plant managers is also

---

41 In this sense, the programme is a combination of a negotiated agreement and informational regulation.

a significant factor and the authors cite factory managers learning more about their own plant's pollution emissions and abatement opportunities as a key impetus for abatement (Afsah *et al.* 2000). In India, even without government support or any available centralised database, one NGO has developed a somewhat similar Green Rating initiative.[42]

Informational regulation is growing rapidly, partly because the success of some of the early initiatives has generated interest in their expansion, partly because it offers a cost-effective and less interventionist alternative to command-and-control in a period of contracting regulatory resources, partly because of its capacity to empower communities and NGOs, and partly because changes in technology make the use of such strategies increasingly viable and cost-effective. The importance of this last factor should not be underestimated. It is becoming easier to track pollution, to detect and trace emissions, and to make the causal links between sources and harms, providing the potential for a new reporting regime 'characterised by consistent and non-redundant reports, electronic reporting, more context to help interpret the data, and a single, multi-media report' (Helms 1999: 27).[43]

Informational regulation can take a number of different forms. Probably the most successful and best known of these is the use of CRTK and pollution inventories. The basis of these policy instruments is to require individual companies to estimate their emissions of specified hazardous substances. This information is then used to compile a publicly available inventory, which can then be interrogated by communities, the media, individuals, environmental groups and other NGOs which can ascertain, for example, the total emission load in a particular geographical area, or the total emissions of particular companies. The latter information, in particular, facilitates comparison of different firms' emissions and can be used to compile a 'league table' identifying both leaders and laggards in terms of toxic emissions. Such benchmarking exercises, facilitated by easy access to the relevant information, enable the shaming of the worst and rewarding of the best companies. The evidence suggests that well-informed communities use this information both to ensure tight enforcement of regulations and to pressure companies to improve their performance even in the absence of regulations. The foremost example of this approach is the US Toxics Release Inventory (TRI), described in Box 7.

Despite the apparent success of the TRI, there are concerns that the data supplied suffers from many limitations, and about how it is presented to the public. There is little, if any, government oversight of the quality of information provided by industry, the rate of non-reporting is substantial, and there is evidence of cross-media transfers (Antweiler

---

42   The Centre for Science and Environment has sought to obtain data directly from companies themselves, have it reviewed by technical experts, and then further assessed though field visits which include discussions with local pollution control boards, local communities, NGOs and the media. Anticipating the resistance of poor performers to participate (good performers have an incentive to co-operate since they will get reputational credit for their efforts), a 'default option' has been established, whereby any company not co-operating is automatically ranked in the lowest bracket. The pulp and paper industry was the first sector to be ranked successfully under this initiative. See Angarwal 1999 and *Green Ratings: The Indian Pulp and Paper Sector*, www.oneworld.org/cse/html/eyou/grp/grp_programmes_pap.htm.

43   The US EPA, under its Reinventing Environmental Information initiative, is committed, in partnership with the states, to implement core data standards and make electronic reporting available in the Agency's major national systems by 2004. See www.epa.gov/rei/about/aptoc. htm.

Background and compliance problem

A series of major chemical accidents, both within the USA and internationally, exposed the need for communities to have access to information about chemicals being manufactured and stored locally, and for the public at large to have access to information about toxic releases locally and nationally. It was felt that, without such information, communities would be powerless to protect themselves or the environment, and that a CRTK was a prerequisite to effective participation in decision-making.

The compliance innovation

The Emergency Planning and Community Right to Know Act, introduced in the United States in 1986, requires companies in certain circumstances to compile an annual inventory of designated hazardous chemicals; and to supply Material Safety Data Sheets to the public. It also requires industry to estimate and report to government emissions of designated toxic chemical releases to air, water or land. This is then compiled into an annual inventory TRI. The basic function of the TRI is to document the release or transfer of selected chemical pollutants to all media as a basis for developing and monitoring the effectiveness of pollution prevention measures or programmes.

Fung and O'Rourke (2000) characterise the US TRI as creating a mechanism called 'Populist Maxi-Min Regulation', which amounts to a type of environmental blacklisting:

> This style of information differs from traditional command-and-control in several ways. First, the major role of public agencies is not to set and enforce standards, but to establish an information rich context for private citizens, interest groups, and firms to solve environmental problems. Second, environmental 'standards' are not determined by expert analysis of acceptable risk, but are effectively set at the level informed citizens will accept. Third, firms adopt pollution prevention and abatement measures in response to a dynamic range of public pressure rather than to formalized agency standards or governmental sanction. Finally, public pressure ruthlessly focuses on the worst polluters—maximum attention to minimum performers—to induce them to adopt more effective environmental practices. TRI has inadvertently set in motion this alternative style of regulation that has in turn dramatically reduced toxics emissions in the United States.

The results

In the USA, there is evidence to suggest that the TRI can: (1) provide community groups with increased insight and political leverage; (2) expose government and regulatory agencies' shortcomings, creating pressure for stricter enforcement; (3) stimulate pollution prevention by sensitising companies to community pressure; (4) lead to the establishment of 'good neighbour' agreements between local communities and companies; (5) improve the quality of public policy debate; (6) lessen the need for environmental regulation via industry commitments to verifiable reduction targets; and (7) directly influence the price of a firm's stock, serving to reward good environmental performers and punish the bad (Hamilton 1993).

Total releases and transfers reported to the US TRI declined by 46% and 36% respectively from 1988 to 1995 and releases reported under the more recently introduced Canadian scheme declined by 36% from 1993 to 1996. The strategy has been hailed by the former administrator of the US EPA, William Reilly (1990), as 'one of the most effective instruments available' for reducing toxic emissions, a view reinforced by his successor, Carol Browner (cited in Orts 1995: 1,227), and many others (Fung and O'Rourke 2000). The North American Free Trade Agreement Commission for Environmental Co-operation, which produces tables ranking North America's worst polluters based on the TRI and comparable Canadian data, claims similar success.*

* Commission for Environmental Co-operation (North America), *Taking Stock* (Montreal: Commission for Environmental Co-operation, www.cec.org/takingstock/index.cfm?varlan=english [published annually])

Box 7 **CRTK and the Toxics Release Inventory**

and Harrison 1999). Nevertheless, the evidence suggests that, overall, the best pollution inventories already provide a very effective stimulus to voluntary action by companies (Fung and O'Rourke 2000). Recognising the potential value of this approach, a number of countries have followed the US example and introduced laws compelling disclosure of pollution and chemical hazard information. Canada and Australia have introduced national pollution inventories and Mexico is contemplating following suit. European jurisdictions, such as the UK, have adopted different types of law that compel government to disclose information on pollution control and chemical hazards following a number of European Community directives (Purdue 1991).[44]

However, not all such inventories and similar instruments will be equally effective and much will depend on their particular design features. While the large majority of assessments of the US TRI are strongly positive, the more recent Canadian scheme has yet to prove its worth, and there is only weak evidence that the latter has been effective in promoting voluntary emissions reductions (Antweiler and Harrison 1999). In Australia, the National Pollutant Inventory (NPI) was so severely weakened by industry-proposed amendments that environmental groups withdrew from the consultation process and the quality of information available under it remains extremely problematic.

A second form of informational regulation is through the practice of corporate reporting on environmental (and on ethical and social) performance (Owen 1996). Such reports can be used by enterprises both as a means of communicating with stakeholders and as a management tool to enhance performance. They may provide opportunities for building goodwill and protecting corporate reputation, overcoming past bad publicity, enhancing product marketing and communicating with employees. However, many of the early environmental reports were far more like public relations exercises than serious attempts to disclose environmental information of value. Even genuine attempts to provide relevant information foundered because of a lack of common standards as to the type of information to be included. These problems were exacerbated by a lack of independent verification.

It will be a considerable period before corporate environmental reporting reaches the same stature as financial reporting. In recognition of this, the Global Reporting Initiative (GRI), established in 1997 and convened by CERES and the United Nations Environment Programme (UNEP), aims to develop globally applicable guidelines that can be used voluntarily by reporting organisations, and enable internal and external stakeholders to gain a better comparative understanding of the environmental credentials of different enterprises. An increasing number of enterprises are also choosing to have their reports externally verified, thereby providing a credibility with external stakeholders that is otherwise lacking.

44 Such directives relate to Freedom of Access to Information and Control of Major Hazards of Industrial Activities (Quality Environmental Management Sub-Committee, President's Commission on Environmental Quality 1993: 9). However, in 2000 the EU Commission adopted plans for the introduction of a new European Pollutant Emission Register (EPER). This register will be a key element of the European Council Directive on Integrated Pollution Prevention and Control (IPPC) to be implemented in 2003. As from that time, member states will be obliged to report to the European Commission every three years on emissions of 50 pollutants from about 20,000 individual industrial facilities covered by the IPPC Directive. Norway has gone somewhat further than most other EU countries, setting up a website offering public access to the environmental profiles of individual businesses.

In the long term, environmental reports that use common reporting criteria and measurements and standardised (possibly sector-specific) formats, and which are independently and professionally verified by third parties, will provide a variety of external stakeholders with valuable information which can be, and no doubt will be, used to reward good performers and to shame recalcitrants into improvement. For example, as with pollution inventories, communities, NGOs, insurers and stock markets are likely to use this information as a means of ranking firms according to their environmental performance and responding accordingly. The growth of socially responsible or 'ethical' investment funds, which seek to balance financial performance with social and environmental issues, may provide additional significant rewards to good environmental performers, as these funds become significant players in investment markets.

A third form of informational regulation is product labelling and certification. Surveys indicate that many consumers are taking environmental considerations into account when they purchase goods and services (Dawson and Gunningham 1996). There is evidence, however, that unassisted markets do not provide accurate information to consumers and in some cases may mislead them about the environmental performance of specific products (Cohen 1991). In order to inform the public about the environmental 'soundness' (or otherwise) of various consumer products, governments can contribute to the development of labelling standards, and of eco-labelling schemes. This can help inform consumers, and sustain markets for environmentally appropriate goods and services (OECD 1991). Private accreditation schemes, with appropriate safeguards, might achieve similar results. However, the experience of establishing eco-labelling schemes within and between nations has been mixed at best.[45] For this reason, some governments have withdrawn from eco-labelling, leaving the development of labelling criteria to non-governmental bodies. Others have confined their involvement to issues such as energy-efficiency labelling of consumer products, where measurement is relatively easy and the results easily disseminated to and understood by consumers.

Finally, it must be emphasised that, like most other policy instruments, informational regulation strategies work better in some circumstances than others. The empirical evidence suggests that they work best with respect to large companies and well-educated

45   With the notable exception of the German Blue Angel scheme, most national broad-based labelling schemes have experienced strong industry opposition, which has greatly limited their coverage and effectiveness. In addition to design difficulties, there are considerable costs associated with the establishment and ongoing operation of labelling schemes, particularly those that attempt to provide a full 'life-cycle analysis' of products (Grodsky 1993). Several schemes intended to become self-funding have in fact required continued government assistance (Dawson and Gunningham 1996). There is also cause to doubt whether the information provided under such schemes is necessarily 'full information' and thus of any use (Menell 1996). The European Union 'flower' scheme, which was intended to provide a single European approach to labelling, has been particularly disappointing. As of December 2000, only around 60 companies had applied to have their goods certified under the scheme, which has had little impact in sectors dominated by a few large companies. Moreover, some of the criteria are so stringent that virtually no firms can apply for them. For example, until recently, no washing machines in the EU market met the standards, while fewer than 1% of light bulbs qualify (ENDS 2000b). The scheme has been described as 'moribund' and EU resources to support it have been substantially reduced (ENDS 2000a). Acknowledging the limitations of the scheme, a revised framework is to be introduced, which will apply to both products and services, but it remains to be seen whether this will be any more successful.

communities (Hartman *et al.* 1997; Afsah and Vincent 1997). CRTK, for example, relies heavily on the energies of local communities in using the information and pressuring enterprises to improve their environmental performance. Where an environmental hazard involves no immediate threat to human health, or where there is no identifiable local community, or where we are dealing with non-point-source pollution, not readily measured and traced back to its origins, then this instrument has far less to offer. Similarly, corporate environmental reporting is dependent on the willingness of public-interest groups to follow through on its results and to both shame bad performers and praise good ones. Finally, eco-labelling relies on the willingness of consumers to buy 'green' products and on their capacity to distinguish between these and other classes of product.

Recognising these limitations, an integrated strategy, using informational regulation in combination with other instrument types, is demonstrably likely to be more effective than a stand-alone approach, as the Canadian experience indicates (see Box 8). This example also serves to illustrate that informational regulation need not be particularly complex, nor introduced at federal level, to be effective. Even state government regulators without the constitutional or political capability to introduce pollution inventories can achieve substantial results through the use of much cheaper and simpler strategies. Extrapolating from the available evidence described above, it would seem likely that even something as simple as an online public register of prosecutions subject to modest procedural precautions (such as not publishing cases subject to appeal)[46] would provide a positive incentive to improved environmental performance at minimal public cost.

Compliance and enforcement

In this final section, we propose to highlight a number of important changes in the architecture of compliance and enforcement. First, contract law is replacing criminal law as the principal enforcement mechanism in a number of circumstances.[47] Take the case of unilateral agreements, negotiated agreements and other voluntary codes. These programmes are often criticised for the fact that they cannot be enforced by means of the traditional criminal law. Industry associations, for example, cannot fine their members or send them to jail; only the state has these powers, and only in cases where specific criminal laws are proved to have been breached. Even where the state is a party to a negotiated agreement it lacks these powers, although it may choose to return defaulters to the traditional regulatory system and its sanctions (as with the Dutch covenants) or introduce new regulation where voluntarism fails.

But the lack of criminal law powers is not the end of the enforcement story. On the contrary, contract law is becoming an important substitute enforcement tool, at least for those agreements that are contractually binding. For example, in 2000 a personal information protection programme under a contract-based voluntary code was enforced

---

46 This practice has been adopted by the UK Health and Safety Executive.

47 These also include regulatory flexibility initiatives in the USA. On the administrative law problems that this may raise, see Freeman 2000.

Background and compliance problem

Previous empirical analyses on compliance, monitoring and enforcement issues have focused their attention strictly on the impact of traditional monitoring (inspections) and enforcement (fines and penalties) practices on the environmental performance of polluters. Other analyses have focused their attention on the impact of public disclosure programmes. But a further compliance issue is whether or not these programmes can create incentives in addition to the incentives normally set in place through traditional means of enforcement such as fines and penalties. Is a public disclosure strategy likely to be a valuable complement or alternative to traditional enforcement and how will it impact on compliance?

The compliance innovation

As a means of bringing extra pressure to bear on non-compliant organisations and in order to bring about greater transparency, the British Columbia Ministry of Environment (MOE), publicly lists firms that either do not comply with the existing regulations or that are of concern to the MOE and where the Ministry continues to undertake legal action for those violating the regulation.

The results

An empirical study of the MOE scheme found that public disclosure is providing reduction incentives beyond traditional compliance levels (i.e. beyond those set in place through traditional regulatory approaches of enforcement and fines and penalties). Indeed, the study suggests that the public disclosure strategy had a larger impact on emission levels and industry compliance than fines and penalties, although stricter standards and heavier penalties would also significantly impact emission levels (Foulon *et al.* 1999).

However, although the evidence suggests that information strategies are useful, they cannot necessarily replace traditional enforcement practices, not least because not all firms will be vulnerable to reputational pressures or will respond as anticipated. The most desirable strategy is a combination of complementary instruments. As the British Columbia study concluded:

> First, the presence of clear and strong standards accompanied with a significant and credible penalty system does send appropriate signals to the regulated community which then responds with a lowering of pollution emissions. Secondly, the public disclosure of information of environmental performance does create *additional* and *strong* incentives for pollution control. These results suggest that both regulation *and* information belong in the regulator's arsenal [emphasis original].

Box 8 *Information disclosure of prosecution and compliance information*

through the courts in precisely this way.[48] Similarly, in 2001, the Forest Stewardship Council (FSC), the central rule-making and rule-oversight body for a sustainable forestry certification programme,[49] suspended the activities of a European-based FSC-accredited certification body, again on the basis of breach of the contract that exists between the FSC and accredited certification bodies (and again between certification bodies and individual forestry operations). To give one further example, by mid-2001, the American Forest

48 See further volcoded-1@synergis.ic.ga.ca posting of 13 March 2001. I am grateful to Kernaghan Webb for this information.

49 www.fscoax.org

and Paper Association, which runs a unilateral programme (i.e. industry self-regulation) had expelled a total of 16 industry members for failure to uphold the standards set by the programme.[50]

This is not to suggest that most voluntary codes and agreements are being enforced effectively by this mechanism. Indeed, the enforcement failings of voluntary approaches are often serious. Some are not, in any event, contractually binding, and many lack formal mechanisms of monitoring and compliance assurance. But, at the very least, a contract offers an important alternative enforcement mechanism where the political will is there to use it. The effectiveness of the contract may also depend on the extent to which there are identifiable motivations for contractors to take action to ensure compliance by contract partners with the terms of voluntary programmes.[51] For example, it has been suggested by Webb that

> the likelihood of publicly disclosed enforcement actions for voluntary code programs will increase where the rule making and oversight body for that program has identifiable, well-represented interests which are different from the industry being regulated (albeit voluntarily regulated) and is distinct from the rule implementation body and indeed separate from the individual adherents (volcoded-1@synergis.ic.ga.ca, posting of 4 April 2001).

A second mechanism which is taking on increasing significance as a vehicle of informal social control is corporate shaming (Braithwaite 1989). This can take a variety of different forms, and can be targeted at either senior or middle management. For example, in relation to INPO (see Box 4 on page 100), Rees describes how CEOs were required to attend an annual meeting at which the performance rankings of all members were disclosed. Those who consistently fell into the bottom group experienced considerable pressure from their fellow CEOs to improve their performance. Since there existed a 'community of shared fate' and all CEOs were aware that they were 'only as strong as their weakest link', there was a considerable incentive to prevent free-riding by this means (Rees 1994). Similarly in relation to Responsible Care, the various leadership groups designed to facilitate sharing of knowledge and experience, serve not only to bring members of different companies together for this purpose, but also to exert peer-group pressure on poor performers (Rees 1997). Research in other areas also confirms 'the importance of an individual's reputation and of informal sanctioning as a means of private governance' (Furger 2000: 7).

Shaming takes place in a different form under various information regulation initiatives described earlier. For example, under the TRI, we saw how companies were required to calculate and disclose estimated emissions of specified hazardous substances. This information is then used by environmental groups and others to develop league tables and similar mechanisms which can then be publicised in order to shame the worst performers. Thus the Environmental Defense Fund (EDF) has developed a publicly accessible database called Scorecard, which is based on TRI data and provides information on pollutants and rankings of individual enterprises. There is no doubt that, notwithstanding methodological criticisms, such databases have succeeded in pressuring large chemical companies and others to address, and in some cases substantially improve, their environmental performance. The US General Accounting Office estimates

---

50 See www.afandpa.org/forestry/sfi/menu.html.
51 See further volcoded-1@synergis.ic.ga. posting of 4 April 2001.

that 'over half of all [TRI] reporting facilities have made one or more operational changes as a consequence of the inventory program' (US General Accounting Office 1991) and EPA credits a 40% reduction in toxic chemical releases to the TRI.[52]

A third shift has been to structure policy in such a way that the market rather than government regulatory agencies will provide the appropriate incentives and surrogate enforcement mechanisms. As we have seen, most economic instruments seek to send appropriate market signals by taxes, charges, subsidies or shifting property rights, rather than by prescribing particular means to obtain particular outcomes and enforcing this through policing and fines. While we have pointed out that such mechanisms are not self-enforcing, nevertheless they rely far less on conventional regulatory monitoring and enforcement than on command-and-control. For example the entire US acid rain programme relies on only a very small regulatory presence.

Another example of this shift towards reliance on the market involves informational regulation. As we have seen, it is not only environmental groups and the general public that react to information about environmental performance but also financial markets. There is evidence that markets respond to information such as TRI figures, by rewarding better environmental performers and punishing the worst, in terms of stock values (Hamilton 1995; Khanna et al. 1998; Konar and Cohen 1997). The precise motivation is not well understood, but it may well be that good environmental management is viewed as a useful indicator of good management generally, and that companies with a good environmental record run a lesser risk of incurring substantial environmental liabilities, such as those imposed under Superfund legislation.

A fourth trend, most evident in regulatory flexibility initiatives and proposals for two-track regulation, is to reward large enterprises for going beyond compliance, and to help them do so by providing them with considerable autonomy and flexibility, as well as other incentives, but subject to certain safeguards. Rather than the state policing and enforcing directly, the latter schemes involve attempts to 'lock in' continuous improvement and cultural change by requiring 'green track' firms to implement an environmental management system, the use of third-party independent auditors rather than government regulators to monitor that system, and transparency and community dialogue requirements which facilitate community and environmental groups also playing a role both in critiquing and monitoring firm performance. Once again, there is evidence of a regulatory reconfiguration and a 'de-centring' of the regulatory state.

Fifth, there is a broader trend towards harnessing a range of second and third parties as surrogate compliance agents. The role of second parties, such as industry associations, and their involvement in co-regulation and industry self-management, is examined in our case studies of ozone depletion and the mining sector. The potential role of third parties, from financial institutions to environmental and other pressure groups, is illustrated by some of the examples given in the preceding paragraphs and by our case study on facilitative regulation in Chapter 8. However, the participation of third parties, particularly commercial third parties, in the regulatory process is unlikely to arise spontaneously, except in a very limited range of circumstances where public and private interests substantially coincide (Gunningham and Rees 1997). There remains, therefore, a significant role for government in facilitating, catalysing and commandeering the participation

---

52   R.B. Outen of Jellinkk, Schwartz and Connolly, Inc., Arlington, VA, 18 October 1999, quoted in Herb and Helms 2000.

of second and third parties to the cause of environmental improvement, as illustrated in Box 9 (see also Gunningham and Grabosky 1998: Ch. 3).

Our broader point, which we made in our previous writing, is that, by expanding the regulatory 'toolbox' to encompass additional players, some of the most serious shortcomings of traditional approaches to compliance can and are being overcome (Gunningham and Grabosky 1998). Third parties are sometimes more potent than government regulators (the threat of a bank to foreclose a loan to a firm with low levels of liquidity is likely to have a far greater impact than any existing government instrument). They are also often perceived as more legitimate (farmers are far more accepting of commercial imperatives to reduce chemical use than they are of any government-mandated requirements). In any event, government resources are necessarily limited, particularly in an era of fiscal constraint. Accordingly, it makes sense for government to reserve its resources for situations where there is no viable alternative but direct regulation. Indeed, the large majority of the next-generation strategies examined in this chapter facilitate this objective.

Conclusion

There are a variety of next-generation policy instruments that, if used judiciously and applied in appropriate circumstances, are likely to achieve far more than the regulatory status quo, both in terms of efficiency and effectiveness. Beyond the specific lessons relating to individual policy instruments discussed above, a number of broader themes emerge.

First, even with the most promising approaches, very close attention must be given to regulatory design. For example, public VAs are likely to be ineffective unless they set environmental performance goals sufficiently high, and include provision for measurement and verification. Similarly, CRTK initiatives, while demonstrably very effective in some circumstances, can fail miserably if they do not control for cross-media transfers and various forms of evasion. Economic instruments, too, have had greatly varying success for this reason, as the experience of environmental taxes and charges (commonly set too low) amply demonstrates. And, while some of the US regulatory flexibility initiatives have so far been disappointing, this may be in large part not because of any inherent flaws in this approach but rather because of the lack of an adequate statutory basis, and other design flaws described above.

Second, instrument combinations usually (but not invariably) work far better than single instruments acting in isolation. For example, unilateral voluntary initiatives have proved seriously flawed as a 'stand-alone' instrument, not least because of the large gap usually found between the public and private interest and the considerable opportunities for free-riding. However, when underpinned by regulatory and, ideally, third-party involvement they can provide industry ownership and engagement, facilitate managerial expertise in ethical and legal compliance, and take some of the weight off overstretched government regulators. Similarly, negotiated agreements have been found to work best when combined with conventional regulation or with economic instruments.

Third, different types of instrument work best in different circumstances. For example, tradable permits can be very effective, but only when applied to circumstances where

Background and compliance problem

For many years, the regulatory regime for the prevention of intentional oil spills (pursuant to an international treaty) was almost wholly ineffective, due in no small part to difficulties of monitoring, and, in some cases, to a lack of either enforcement resources or political will on the part of member countries. Furthermore, in the absence of government intervention and the imposition of penalties, third parties had no incentives to contribute significantly to the reduction of oil spills.

The compliance innovation

However, all this changed when a new regime was introduced under the International Convention for Prevention of Pollution by Ships (MARPOL), requiring tankers to be equipped with segregated ballast tanks (SBTs). The new regime facilitated initial surveys and inspections by non-governmental classification societies, and consequently made it hard for a tanker lacking the required equipment to receive the classification and insurance papers needed to trade internationally.

The results

The new equipment standards requiring tankers to install SBTs achieved almost 100% compliance, despite the fact that this entailed significant investments with no offsetting benefits, at a time when oil prices were falling and creating pressure to cut costs. While other provisions of the Convention were frequently violated (such as MARPOL discharge standards limiting amount and location of discharge), tanker owners installed SBTs. Their action was not in response to the threat of enforcement by other countries but rather as a direct response to a 'coerced compliance' approach that was aimed primarily at preventing violations. The new regime facilitated coerced compliance by three powerful third parties, namely non-state classification parties, ship insurers and shipbuilders. None of these parties had any interest in avoiding the new regime, yet shipowners were critically dependent on each of them. Together, and in conjunction with state action, they achieved far more than state action alone was ever likely to. R.B. Mitchell argues that clearly tanker captains still have incentives to discharge waste oil at sea, but successful waste regulations utilising the influence of third parties have prevented their acting on those incentives:

> At the most general level, treaty provisions alter behaviour when they create the conditions for a strategic triangle of compliance in which agents have the political and material incentives, the practical ability, and the legal authority to undertake (or refrain from) a specified activity . . . Tanker owners for instance, always had the ability to install SBT equipment, but MARPOL gave them a new incentive to do so: Their ability to trade internationally was now contingent on compliance with MARPOL equipment requirements, as determined by classification societies and insurance companies.

Box 9 *Lessons from intentional oil pollution*

Source: Mitchell 1994

they can be easily monitored and verified. Negotiated agreements also work much better in some circumstances than in others. Key success factors have been identified as: (1) the degree of mutual respect and trust between the participants; (2) whether real alternative policies, such as regulation, can be pursued in the absence of an agreement; (3) the 'accessibility' of the target group, based on factors such as its homogeneity, the number of players, or the existence of a strong representative body able to negotiate on behalf of the group; and (4) the existence of environmental competitive advantages in the target sector (De Clercq *et al.* undated). Again, it is only a minority of circumstances that lend themselves to co-regulation or industry self-management, usually when industry has a strong self-interest (for example, what Rees calls a 'community of shared fate') in working co-operatively to improve its actual environmental performance.

Fourth, the progress of regulatory reform is a dynamic and adaptive one. We can learn from the mistakes of each generation of instruments to better design the next one. So, while the first generation of voluntary initiatives has a very mixed track record, there is an opportunity to achieve far greater success for the next one by learning the lessons of past mistakes. For example, policy-makers have learned the importance of reducing the role of industry in the target-setting process, of curbing free-riding, of building in credible regulatory threats, of ensuring that commitments are enforceable, transparent and capable of being monitored. They have, in short, identified a number of 'success' criteria which, if followed, may achieve more positive results for the next generation of instruments.

Finally, notwithstanding the increasing importance of these new-generation enforcement mechanisms described above, there will still be a need for an underpinning of enforcement under traditional command-and-control regulation. Few, if any, alternative enforcement mechanisms work optimally in isolation. Not all free-riders will be curbed effectively by contract, and the role of the state in reinforcing even the best of voluntary initiatives is graphically illustrated by Rees's study of INPO (Box 4 on page 100). Regulatory flexibility initiatives will encounter firms that seek to gain the benefits of 'green track' regulation while trying to avoid the costs. Only if such firms can be detected and returned to a traditional enforcement track can a level playing field, and the integrity of such initiatives, be preserved. Economic instruments, too, require enforcement, without which it may be tempting for firms to disguise their level of emissions, to cheat in relation to tradable permits, and otherwise manipulate the rules. Even informational regulation initiatives, such as the TRI, will work better when government takes steps to monitor the quality of information being provided, and to prosecute for wilful misrepresentation (see also Box 8 on page 128). In essence, regulatory reconfiguration does not imply the removal of the regulatory state, but rather its repositioning.

7
VOLUNTARY APPROACHES TO ENVIRONMENTAL PROTECTION
Lessons from the mining sector*

As we saw in the previous chapter, environmental policy-makers are increasingly adopting voluntary initiatives to supplement, complement or replace direct government regulation. From a government perspective, the reasons for this interest in voluntarism include the limits of command-and-control regulation, the need to compensate for increasingly inadequate regulatory resources, the benefits of promoting dialogue with and raising the environmental awareness of the private sector, and the attractions of generating beyond-compliance outcomes (for a detailed analysis, see Moffet and Bregha 1999).

For industry, the benefits include the avoidance of more direct regulation, greater flexibility in reaching targets, risk management (including protecting themselves from potential litigation) and reputation assurance (particularly for industry leaders). This last factor has become of critical importance to large, highly visible transnational corporations, which increasingly recognise the need to manage their relationships and maintain their credibility not only with governments but also with a broader range of stakeholders, including communities, NGOs and workers.[1]

Although voluntary initiatives have proliferated across a range of issues and in a variety of countries, we still know very little about their effectiveness or about how best to design them to achieve optimum efficiency and effectiveness. The empirical literature is very limited, due largely to the recent introduction of this approach and the lack of data collection and reporting requirements in many such initiatives (Davies and Mazurek 1996; NRC 1997; Beardsley 1996; EEA 1997; Harrison 2001; OECD 1999). Even such evaluation as has taken place has been of doubtful credibility (see Chapter 6).

* The authors are grateful to Jon Higgins for his detailed comments on an earlier draft.
1 As such, 'voluntary' initiatives may owe far more to a calculated response to external (and internal) pressures than they do to any spontaneous wish to become more socially or environmentally responsible. Accordingly, this chapter recognises that voluntarism is a question of degree, not an absolute, and embraces within it initiatives that may in significant part be responses to external forces.

Against this backdrop of a paucity of reliable empirical evidence, this chapter examines the experience of one particular form of voluntary initiative—industry self-regulation—in relation to the mining industry, whose activities can have profound environmental and resource implications. As we will see, an industry-specific approach is valuable because the appropriateness of voluntary initiatives, and the design features necessary to maximise their chances of success, are context-specific. That is, there is no single approach that is likely to work in all industries or in all circumstances.

The remainder of the chapter is organised as follows. We first provide a context by elaborating on what we believe to be the centrally important concept of social licence as a driver of corporate voluntarism. We then explore the mining industry experience (comparing this to the similar but empirically more advanced experience of the chemicals industry). Finally, we seek to draw some broader lessons about the design and appropriateness of voluntary approaches as a policy instrument, and some general conclusions.

A context: the importance of social licence

We can get a better understanding of the role of voluntary approaches, particularly industry-specific regulation, if we first ask *why* industry is increasingly attracted to the use of such instruments, and what purpose(s) they would serve. The main impetus for the introduction of such approaches (mainly in the form of industry codes of practice) is the need for the industry, or at least the reputation-sensitive industry, to maintain its environmental credibility (and often in so doing, pre-empt the threat of tougher legislation; see e.g. Wells 2001). For example, in the case of the mining industry, as one commentator has pointed out:

> worldwide, mining is faced with a pattern of low credibility and social opposition, which derives from a general perception that mining is a dirty business. Mining is seen as inherently destructive, in that it destroys the environment, and leaves nothing positive behind when it packs up and goes. The image of abandoned mines, tailings dumps, waste-rock piles, and abandoned communities has significant resonance with the general public (*Mining Journal* 1999: 441).

The problems are usually greatest in developing countries where mining enterprises commonly confront a legacy of conflict, struggles over the distribution of the benefits of mining, legislative inconsistencies generated by a variety of different reform processes, and a perceived lack of legitimacy in the laws and regulations on which foreign companies rely (Joyee and Thomson 1999, cited in *Mining Journal* 1999). This last problem may be particularly serious given that there are commonly unresolved problems of legitimacy and transparency related to the entire process of mineral resource development and that a transnational company may be seen as aligned with a government that lacks legitimacy in rural areas. And even where there is the political will to regulate the environmental impact of the mining sector, governments and regulators commonly lack the capacity to do so.

As a result, the mining industry faces an urgent need to gain and maintain legitimacy and social acceptance. It cannot rely merely on the fact that it claims to be in compliance with local environmental laws to achieve this. It is particularly vulnerable to criticisms from a combination of local and international NGOs. These groups, benefiting from the global revolution in communications and information technology, not only have far greater knowledge of mining operations than previously but can disseminate that information rapidly, and in forms (e.g. digital photography and the Internet) that facilitate highly sophisticated media campaigns directed to individual corporations or to the industry at large. The Brent Spar saga, albeit in another resource sector, is a dramatic illustration of the impact that sophisticated NGO media campaigns can have on corporate reputation and profits. The environmental and social damage caused by the Ok Tedi mine in Papua New Guinea at one stage threatened to become a comparable media disaster for its owners, at least at the regional level.

An enterprise that builds its reputation capital and social licence can also turn this into a competitive advantage: 'Reputation capital represents a communications bridge which predisposes NGOs, communities and other groups to enter into open discussion rather than hostile opposition. Reputation capital carries with it credibility, such that the upfront costs and risk associated with gaining social acceptability are reduced' (Joyee and Thomson 1999, cited in *Mining Journal* 1999). Those with reputational capital will be those that benefit from greater access to government and planning approvals, community acceptance and preferred access to prospective areas and projects. As one major mining enterprise recently pointed out:

> for mining companies, the critical issue for long term prosperity is access to the resource base. Those companies that can demonstrate a high standard of environmental and social performance will be well placed in securing access to resources. It will also help companies to better manage risk and community relations and enhance their reputation (WWF–Placer Dome Asia Pacific 2001: 1).

Against this background, how can industry convince society that it is acting responsibly in the way it exploits resources, and that it is doing so in a manner that is compatible with the concept of sustainable development? How can individual companies demonstrate that they are responsible environmental actors that can be trusted to mine in a particular area in a developing country without poisoning the local rivers, irreparably damaging the local environment and destroying the culture of indigenous peoples? How can companies avoid, for example, more serious accidents involving cyanide such as the Baia Mare disaster in Romania, the Kumtor incident in Kyrgyzstan and at the Ok Tedi mine in Papua New Guinea? Put more broadly, how can a company, and the industry as a whole, protect its 'social licence to operate'?[2]

An important distinction here is between action that is required to protect the reputation of an individual company, and action that is needed to protect the reputation of an entire industry. Of course, there is nothing to prevent individual companies from improving their own environmental performance without adopting a particular code of practice, although they may gain greater credibility (to the extent that such a code is respected by external stakeholders) by doing so.

Incentives for such individual action include not only the protection of reputational capital but also competitive advantage and increased profitability, to the extent that they

---

2   Details of the cyanide code of conduct can be found at www.cyanidecode.org.

can either: (1) identify win–win solutions which, for example, enable them to save substantial sums of money through more efficient use of resources; or (2) gain a market advantage, to the extent that purchasers are willing to exercise a preference for goods produced in an environmentally friendly (or less unfriendly) manner. However, better environmental performers have 'largely gone unrecognised and unrewarded by the market and the public because of the absence of a credible mechanism that can differentiate companies on the basis of their environmental and social performance' (WWF–Placer Dome Asia Pacific 2001: 1).[3]

In any event, it is clear that individual initiatives will not be sufficient to protect the reputation of an industry as a whole, and that, unless the industry as a whole is trusted, then the prospects of individual companies within it may be threatened. This is because a major environmental incident involving an individual company commonly tarnishes the reputation of the entire industry, exposing it to the risk of tougher regulatory requirements, obstacles to development and community backlash. As one industry spokesman put it: 'Businesses can only survive while they have society's acceptance for their activities. Once that acceptance is lost, there is only one way to go' (Holmes 1992: 3). The need for collective action is greatest in the case of industries that rely on intermediaries to market their products directly to consumers, such as chemicals, oil, wood pulp, textiles and mining. As Nash (2000: 7) points out: 'industries that manufacture commodities that require further processing before sale to end-users tend to assume a collective identity in the public's mind. The problems of one company color public perception of the entire industry.'

What this means in practical terms is that each company in such industries must act as its brother's keeper. Thus a mechanism must be found, nationally and internationally, that enables the industry to continuously improve the environmental performance of all companies, large and small. Such a mechanism must be capable of improving the industry's poor public image,[4] restoring public faith in the industry's integrity and taking the heat out of demands for stricter government regulation.[5]

In the case of the mining sector the challenge is a considerable one. It has been pointed out that mining companies compete globally for the most productive minerals and metals deposits, which are attributed to them in the form of operation licences by public authorities, and that 'in recent years, companies' environmental performances have become more and more important for the attribution of these licences. Thus environmental performances are a strategic issue for individual mining companies and they do not necessarily gain from collective action' (Bomsel et al. 1996: 85).

---

3   Whether independent certification (as under the model of the Forest Stewardship Council or the Marine Stewardship Council) would provide a mechanism that allows minerals and metals from well-managed mines to access environmentally sensitive markets is uncertain and untested.

4   This was acknowledged by ACIC (Australian Chemical Industry Council, as it then was) former Chief Executive, Frank Phillips, who said that the plan was developed in response to the industry's poor public image. See Smithers 1989.

5   As former Canadian Chemical Producers' Association President Jean Belanger put it: 'if we could figure out a way of becoming proactive, then we could lessen demands for that degree of regulation' (Mullin 1992: 28).

The mining industry

The task of protecting a social licence is multi-dimensional as it involves operating at several levels and engaging in dialogue with a variety of stakeholders in order to gain credibility and legitimacy. One partial but important strategy for gaining social licence *at an industry level* is the development and implementation of industry codes of practice and other voluntary approaches that establish industry standards of environmental performance.

Industry groups in a number of countries (most notably Canada and Australia) have to varying degrees contemplated or introduced such codes. Most notably, the International Council on Metals and the Environment has an Environmental Charter and an Environment code to which members must commit. This consists of a general set of environmental, product stewardship and community responsibility principles which focus on management and organisation, and which stipulate the sharing of information on good environmental practices among members. Among the obligations specified are to 'meet all applicable environmental laws and regulations and, in jurisdictions where these are absent or inadequate, apply cost–benefit management practices to advance environmental protection and to minimize environmental risks'. However, there are no significant incentives to join; neither is there independent monitoring, or sanctions for non-compliance. Unsurprisingly, there is very little evidence that this initiative has made a significant contribution to environmental protection, and a formal analysis by Bomsel *et al.* (1996) concludes that 'firms do not feel constrained to improve their environmental performance by the ICME charter'.[6]

Also of significance is the Global Mining Initiative, involving a number of major mining enterprises, which has commissioned a study seeking to analyse the factors that could help the mining sector better contribute to sustainable development, and which aims to reach an understanding with all stakeholders, and the industry's critics, on the role that mining should play in the transition to sustainable development.[7] However, at the time of writing this initiative is still in its very early stages.[8]

A much more advanced example of what a mining industry code might involve is the Australian Minerals Industry Code for Environmental Management (AMICEM), which was launched in 1996 and substantially revised in February 2000. This has been described by UNEP as 'one of the most comprehensive voluntary codes yet devised for the mining industry, and the only code to require the disclosure of environmental performance' (Gould 2000). For these reasons, it has been proposed as the model for mining industry self-regulation internationally.

While the precise motivations of the industry in establishing the code remain a matter of speculation,[9] they certainly included a concern to protect the reputation of the Australian mining sector as a whole, and, with it, the reputation of individual companies and

---

6   The ICME is currently being replaced by a more globally oriented organisation that will represent all sections of the mining industry, to be called the International Council on Metals and Mining.

7   *Mining, Minerals and Sustainable Development*, www.iied.org/mmsd.

8   See www.globalmining.com.

9   The MCA has stated that the Code was developed 'to demonstrate its commitment to continual improvement in environmental management, and to be open and transparent in its dealings with the community' (MCA 2000).

their capacity to gain access to markets internationally. Signatories to the code commit to:

⚫ Integration of environmental, social and economic considerations into decision-making and management, consistent with the objectives of sustainable development

⚫ Openness, transparency and improved accountability through public environmental reporting and engagement with the community

⚫ Compliance with statutory requirements as a minimum

⚫ A continually improving standard of environmental performance and, through leadership, the pursuit of environmental excellence throughout the Australian minerals industry

Obligations under the code are: progressive implementation; production of an annual public environment report within two years of registration; completion of an annual code implementation survey to assess progress against towards the implementation of code principles; and verification of the survey results, by an accredited auditor, at least once every three years. The code also requires implementation of an environmental management system. In its most recent manifestation it also seeks to create mechanisms to foster the exchange of information, experience and 'lessons learned'. In June 2001 there were 41 signatories to the code, representing over 90% of the industry's mineral production. While this figure, supplied by the Minerals Council of Australia (MCA), suggests that the code has been widely adopted, it may disguise the extent to which smaller companies have so far declined to participate.[10]

The Australian code and other mining industry initiatives are of too recent origin to be the subject of detailed empirical evaluation at this stage[11] (although the World Wide Fund for Nature [WWF] has produced an environmental scorecard based on a desktop analysis, which rates report content but not performance; see Box 10).

We do have much greater empirical experience of other, somewhat similar, codes of practice.[12] In particular, since the code bears considerable similarities to, and was modelled substantially on, the chemical industry's much more developed Responsible Care programme, we can at the very least extrapolate from the experience of Responsible Care

10 The above figures were provided on request by the MCA, which did not supply the number of non-participants. Those SMEs that are participants tend to be those mining in highly sensitive areas such as major tourist regions, or those mining directly under a township.

11 At the time of writing the Mining Minerals and Sustainable Development Project, through its Australian component, is examining whether a company's status as a signatory to the code is correlated with behaviour, as reflected by reported initiatives relating to sustainability. See Sinclair *et al.* 2001.

12 The evidence suggests that such codes are rarely effective in achieving compliance (i.e. obedience by the target population[s] with regulation[s])—at least if used as a stand-alone strategy without sanctions. This is because self-regulatory standards are often weak, enforcement is commonly ineffective and punishment is secret and mild. Moreover, self-regulation commonly lacks many of the virtues of typically conventional state regulation, 'in terms of visibility, credibility, accountability, compulsory application to all, greater likelihood of rigorous standards being developed, cost spreading, and availability of a range of sanctions' (Webb and Morrison 1996). For a recent and comprehensive survey, see Priest 1998–99.

THE WWF'S SECOND ANNUAL 'SCORECARD' OF ENVIRONMENTAL REPORTS produced by signatories of the MCA Code for environmental management assessed 32 such reports. In broad terms:

- The scorecard results show that there is a wide range in the quality of environmental reports as assessed against WWF criteria.

- None of the reports received a 'pass' mark for external verification. While many of the reports contained a statement from an external auditor, few of these statements are enlightening or provide sufficient commentary on the company's environmental and social performance.

- Some progress has been made with community participation, with nine reports receiving a pass mark compared with none in 1999.

- Many companies continue to ignore the importance of setting and reporting on environmental performance targets, with 13 reports having no targets at all.

- Most of the companies that scored well for environmental performance targets also received a high score for data and for environmental management, identifying those companies that have incorporated environmental management into all levels of the company's activities.

- Despite a review of the MCA code since the 1999 scorecard was produced, there has been very little improvement in the quality of the environmental reports as assessed against WWF's criteria.

Box 10 *Mining environmental reports: WWF second scorecard on environmental reports*
Source: WWF 2000

to evaluate AMICEM in terms of key criteria such as its coverage and ambitiousness, monitoring, sanctions for non-compliance, transparency and credibility. The two industry sectors have considerable similarities in that both have a substantial and high-profile environmental impact, both have had a very poor public environmental image, and both need to protect their reputation capital in order to maintain access to development opportunities across a diverse range of countries, to ward off more interventionist government regulation, and to maintain credibility with external stakeholders.[13]

Responsible Care evolved in the aftermath of the chemical industry disaster at Bhopal, India, in 1984, at a time when the chemical industry internationally faced a serious credibility problem and feared draconian government regulation and serious public opposition to many of the industry's activities. Responsible Care is a self-regulatory programme intended to reduce chemical accidents and pollution, to build industry credibil-

13 However, there are also some significant differences. For example, the chemical manufacturers' associations are in a stronger position than most such bodies to exert pressure for environmental improvement, in part because the industry's characteristics facilitate the development of 'social capital': the development of 'the features of social organisation, such as networks, norms and trust, that facilitate coordination and cooperation for mutual benefit'. As Rees (1997) has demonstrated, the industry is an incestuous one in which companies constantly deal with each other. Strategic alliances, product swapping and technology transfers are the norm rather than the exception.

ity through improved performance and increased communication and to involve the community in decision-making. It is built around a series of industry codes of practice and greater levels of public disclosure and participation with administration by chemical industry associations at national level. The relevant associations rely largely on promulgating norms of industrial conduct, peer pressure, technical assistance and transfer, data collection and self-reporting by members to institutionalise responsibility and ensure compliance. Expulsion of a member for non-compliance is extremely rare.

At best, over 15 years since its inception, Responsible Care has achieved only very modest success. In the United States, there is no publicly accessible aggregate data on enterprises' compliance with industry standards, and informal estimates are that some 30% of members have been 'recalcitrant' in adopting the programme (Sabel et al. 2000). One recent evaluation found that: enterprises that are more greatly influenced by the industry's reputation will more frequently participate in Responsible Care; companies with weaker environmental performance relative to their sectors were more likely to participate in Responsible Care; there is no evidence that Responsible Care has positively influenced the rate of improvement among its members; there is evidence that members of Responsible Care are improving their relative environmental performance more slowly than non-members (King and Lenox 2000).

Overall conclusions from this research are that, because Responsible Care has operated without explicit sanctions for malfeasance, 'it has fallen victim to enough opportunism that it includes a disproportionate number of poor performers, and its members do not improve faster than non-members. Thus the institutional pressure that Responsible Care exerts on its members appears to have inadequately counteracted opportunism' (King and Lenox 2000).

Another study of 16 enterprises found that the Responsible Care programme, in addition to improving community relations and oversight of distributors generally, also changed dramatically the way of thinking of three of the firms, and was a useful and important safety, health and environment tool in another three. However, in ten of the firms Responsible Care primarily helped relations with external constituencies without significantly changing internal behaviour (Metzenbaum 2000; Nash 2000). Indeed, some enterprises (which see environmental practices as marginal to their strategic and competitive objectives) appear to treat Responsible Care as a tool for external image manipulation rather than for genuine environmental improvement (Coglianese and Nash 2001).

However, Responsible Care continues to evolve, and, as previous analyses point out, substantial modifications, including independent third-party audit (already in the process of being introduced in some jurisdictions) and expanded roles for accountability, transparency and consultation, may go a substantial way to improving both its credibility and its capability to deliver positive environmental outcomes (Gunningham and Rees 1999; Gunningham and Grabosky 1998: Ch. 4). For example, Nash points out that in the USA all members are now expected to have implemented fully all management practices and the board has decided to disclose the name of enterprises whose Responsible Care programmes are lagging, to the board's Responsible Care committee (but not to the public). And in 2000, the board decided to rank some aspects of members' code performance and distribute this ranking to the membership (but, again, not to the public) (Nash 2000: 11; see also Howard et al. 1999).

Responsible Care may also have achieved much more than the above analyses give credit for in terms of soft effects (which are rarely dealt with because they are difficult to measure).[14] Rees has demonstrated that the chemical industry associations, through Responsible Care, have facilitated the development of trust among their members, creating an environment within which people work together, share information, provide mutual aid and establish policy. Tangible manifestations of this include Responsible Care's leadership groups,[15] workshops, a mutual assistance network and implementation guides. As a result, 'by increasing interpersonal trust and reducing uncertainty, the development of community lowers transaction costs and makes collective action easier' (Rees 1997).

More broadly, Responsible Care has enabled the development of an industrial morality, a set of norms that generate a sense of obligation, emphasising particular values and structuring choice. Such a morality provides:

> a form of moral discourse capable of challenging conventional industry practices—'This is the way we always do business around here'—including the economic assumptions underlying many of those taken-for-granted policies and practices. In this way, an industrial morality . . . legitimises aspirations other than profit as a good reason for action. It establishes an alternative moral vocabulary, a rhetoric of organisational motive that competes with (and critiques) the native tongue of the business organisation, the language of profits and losses (Gunningham and Rees 1997: 376).

Within such a context, there is also considerable scope for peer-group pressure to act as an effective driver of corporate change. For example, the leadership groups in particular fulfil this role, bringing together representatives of a number of enterprises to share their experiences, their progress and, by implication, their lack of progress.[16]

Similarly, there is the potential for Responsible Care to act as a vehicle for corporate shaming[17] through the spotlight of public exposure of a polluter's moral failings. Certainly the performance indicators and verification mechanisms currently being adopted under Responsible Care could form the basis for identifying recalcitrants and exposing them to the glare of adverse publicity. There is also some anecdotal evidence that to a modest extent such shaming already takes place through the leadership groups. In these ways, Responsible Care provides a vehicle for informal social control: regulation from the inside ('moralising social control') (Braithwaite 1989), rather than regulation from the outside (based on external constraint).

Overall, the various Responsible Care mechanisms designed to develop mutual trust among competitors, to facilitate mutual aid, information and technology sharing, peer support, pressure for corporate shaming and dialogue with local communities, the public and governments, create a climate that can motivate and drive corporate executives to do far more in terms of environmental performance than the law could credibly require. However, despite its considerable potential and strengths, there are also many obstacles

---

14 The following paragraphs are taken from Gunningham and Grabosky 1998: Ch. 4.

15 These groups usually meet quarterly with peers to review progress and to provide and receive assistance. They are reputedly a highly effective way of creating peer pressure, and of enlisting corporate leaders to the cause of Responsible Care.

16 See Posner 1992, citing how such a process takes place during meetings of company chief executives.

17 There is a criminological literature that argues persuasively the importance of a moral dimension to corporate (and individual) behaviour, and documents the considerable extent to which corporations can be 'shamed' into doing the right thing (Braithwaite 1989).

to the success of Responsible Care. Of these, the largest is that environmental protection and private profit do not necessarily coincide, and are not perceived to coincide, particularly given the emphasis of most corporations on short-term profitability.[18] For both corporations and individual managers, the essential dilemma is that they will be judged essentially on short-term performance, and, if they cannot demonstrate tangible economic success in the here and now, there may be no longer term to look forward to.[19]

There may be a number of particular lessons for the mining industry to be learned from the Responsible Care experience. First, to the extent that the code relies on self-reporting as the principal means of monitoring, it will lack credibility with external stakeholders and the public in general. When the industry associations responsible for administering Responsible Care announced yearly compliance figures based on their member companies 'ticking the boxes' and returning questionnaires, these statistics were greeted with great scepticism by external audiences, and as tantamount to students grading their own exam papers. Only very belatedly, and only in a very small number of countries, is Responsible Care turning to external verification and independent audit, as a means of providing credible monitoring and reporting. The leader in this respect is Canada, where an external team comprising two industry and two non-industry representatives (one from the local community) conduct such audits.

Significantly, the most recent version of AMICEM requires verification of the survey results, by an accredited auditor, at least every three years. However, since the code requirements are essentially process- rather than outcome-based (a matter to which we return below), this does not imply independent verification that any particular level of environmental performance is being achieved, but rather that the systems that companies claim are in place are indeed in place.

Second, if the industry association is unwilling to impose credible sanctions for non-compliance with the code, this substantially reduces its credibility. Environmental groups constantly ask for evidence that the relevant industry association is willing to take action to sanction renegade enterprises for non-compliance. In most countries there are none, leading its critics to suggest that Responsible Care tolerates free-riders and lacks the political will or the means to pull up the recalcitrants to the standards set out in its codes of practice. It remains to be seen whether the MCA will adopt a more aggressive approach to sanctions. At present, the position taken is that the code 'is not there to judge

---

18 Because corporations are judged by markets, investors and others principally on short-term performance, they have difficulty justifying investment in environmentally benign technologies which may make good economic sense in the long term, but rarely have an immediate or medium-term payoff. Most areas of reform, including stopping harmful emissions to land, water and air, replacing harmful chemicals with more expensive ones, and cleaning up contaminated land, are vulnerable to these short-term market pressures.

19 Jackall (1988) found that short-term issues overwhelm long-term considerations. In Jackall's view, 'Managers think in the short term because they are evaluated both by their supervisors and peers on their short term results.' As one manager put it: 'Our horizon is today's lunch' (Jackall 1988: 84). Jackall also found that staff mobility, both within and between corporations (often the result of CEO-inspired re-organisations), meant that those who currently occupy a managerial post might feel no urgency about the environmental consequences of their decisions. This was because the threat of immediate governmental retribution, via the EPA, was most unlikely, and the delays in processing environmental actions through the courts meant that, by the time a case was heard, the present incumbents would have moved on, leaving others to deal with the legacy of those decisions (Rogers Jr 1992: 31).

how companies perform, and has no capacity to apply punitive measures when they fail to measure up' (Gould 2000: 22).[20]

Third, transparency will be a critical feature of a code's credibility with the general public. Responsible Care in the USA initially used the slogan 'don't trust us, track us' as a means of demonstrating to the public a commitment to openness and to full disclosure. However, in practice, its members have been loath to make such disclosure. Even now, there is no requirement to make compliance audits public in the United States and the Chemical Manufacturers' Association (now the American Chemistry Council) leaves it to individual member companies to define what constitutes 'full implementation' for their own circumstances (Harrison 2001). The code development and implementation processes are even less open in Europe. Only in Canada and Australia are firms required to publicly report their discharges beyond the requirements of law. In these jurisdictions public involvement is facilitated via a national advisory panel and facility-level committees.

Again, it is too early to judge AMICEM in terms of transparency. Certainly 'signatory values' include 'openness, transparency and improved accountability through public environmental reporting', but no prescriptive environmental reporting requirements are included and only the aggregated progress of code signatories will be made public by MCA, rather than individual public surveys (WWF–Placer Dome Asia Pacific 2001). The early signals in relation to transparency are not encouraging. The mining industry took a leading role in opposing the implementation of a national pollutant inventory (a watered-down version of the US TRI), and was successful in having removed from the NPI proposed reporting requirements in relation to issues such as tailings dams, which the mining industry would prefer not to be the subject of public scrutiny. Another response to its anticipated public environmental reporting initiative has, according to some regulators, been to routinely defend and if necessary appeal all prosecutions, even those where (according to regulators) it would in the past have pleaded guilty. Since the industry can massively outspend local regulatory authorities in the courtroom, this strategy may well be successful in deterring EPA prosecutions and in preserving a clean record, but says nothing about the industry's commitment to improved environmental performance.

Finally, the Responsible Care codes, like ISO 14001, focus on systems rather than outcome-based standards. As such, they leave the setting of goals to individual participants, focusing only on the processes that are in place to ensure that goals are achieved. As Harrison (2001: 231) points out:

> at its core Responsible Care remains a management system. Whether or not it is effective as such, it cannot be viewed as a substitute for goal oriented public policy. For instance, the manufacturing code of practice commits participants to 'be aware of all effluents and emissions to the environment, monitor those for which it is necessary, and implement plans for there control when necessary' . . . It is striking that no specifics are provided about which substances should be monitored, what to report to the community, when control is 'necessary', or how firms should go in 'responding to community concerns'.

---

20   However, minerals industry respondents indicate informally that two (undisclosed) companies have been requested to resign for non-compliance.

The same comment applies equally to AMICEM. WWF has further pointed to the lack of an entry standard or deadlines for improved environmental performance as a related, and equally important, limitation. Their general conclusion is that 'such a wide spectrum of environmental and social performance conforms with the Code—from the very poor level of performance to industry leaders—that stakeholders such as WWF are unable to use compliance with the Code as assurance that an acceptable level of environmental performance is being met' (WWF 2000: 4).

Clearly, the most fundamental issue of all is ensuring that the Code addresses performance as well as process. However, assessing performance is not easy, nor is comparing the performance of different mine sites. Solomon's research suggests that 'there is clearly a tradeoff . . . the greater the focus on outcomes and performance, the greater the need for context and judgment and the acknowledgement of subjectivity . . . metrics of performance are required. However, their combination into aggregated assessments cannot be automatic. The mix of metrics will depend on the audience for verification and its information needs' (Solomon 2000: 95-96). The fact that the verifier will be paid by the audited entity also raises questions of conflict of interest and suggests the need to 'verify the verifiers': for example, via external stakeholder participation, along the model adopted in Canada in respect of Responsible Care.

Broader lessons regarding voluntary initiatives

What are the broader lessons we have so far learned about the design and appropriateness of voluntary instruments as a policy mechanism?[21] This issue can be dealt with under two headings, addressing two discrete questions. First, what internal characteristics are most likely to make voluntary initiatives effective? Second, how can voluntary initiatives be linked with other policy instruments or external pressures in order to increase their effectiveness?

Internal design features

Our limited experience with voluntary initiatives suggests the importance of structuring them in ways that maximise their chance of success. Here, building on previous work by Gunningham and Rees, a number of features can be identified as of particular importance.[22]

Environmental targets

Not all voluntary initiatives involve clearly defined targets; indeed, most do not. The case for more generalised agreements is often that concrete targets are impossible to achieve

---

21 For a recent OECD analysis on related issues, see OECD 2000b. See also UNEP 1992, which identified five criteria as important in making voluntary codes effective: commitment, content, collaboration, check and communicate.

22 See generally Gunningham and Rees 1997; Moffet and Bregha 1992; Lyon and Maxwell 1999, and references therein.

in the early stages and that it is better for participants to feel their way, rather than resisting (and perhaps refusing to enter) an agreement that might commit them to non-attainable targets, or ones that, in retrospect, it is uneconomic to achieve. Far better in these circumstances to at least begin with good-faith obligations of a general nature, and process-based obligations (for example, in terms of developing and implementing an environmental management system). However, in the case of mature agreements (such as the Dutch environmental covenants; see Chapter 6), and those capable of lending themselves to specific quantifiable targets, the adoption of such targets, both for individual firms and across the entire industry sector, is highly desirable. Without them, there is the risk that the initiative may become vacuous, degenerate into 'greenwash', and lose credibility. Responsible Care and AMICEM are both vulnerable in this respect.

Accountability and transparency[23]

Those who are held accountable under an agreement know they must explain and justify any questionable actions. This tends to both discipline and constrain decision-making. But how can accountability best be achieved? One of the principal mechanisms by which accountability can be fostered is transparency. Arguably the first step towards transparency is the public announcement of the principles and practices that participants accept as a basis for evaluating and criticising their performance. When first promulgated, these norms are often stated in very general terms, but can later be refined into detailed codes of management practice. The important point here is how a participant, by clarifying the standards it sets for itself, including performance indicators and implementation timetables, also provides more precisely defined measures for evaluating and criticising its performance. With increasing transparency, in short, accountability is more readily maintained—even more so when environmental groups begin to develop more sophisticated measurement tools and environmental scorecards, such as the WWF has done on mining environmental reports.

The next critical step towards achieving transparency is the development of an information system for collecting data on the progress of implementing the agreement. The process usually divides into two parts: (1) reporting and data collection; and (2) collation and analysis of data. Self-reporting is the most common form but raises concerns of conflict of interest: firms may be tempted to be less than frank where full disclosure would reflect poorly on their performance. And what about enterprises that are unwilling or unable to respond fully to often cumbersome reporting requirements? This brings us to verification and monitoring.

Monitoring and verification

The third and final step in achieving transparency—monitoring performance—also seems to be the most demanding and controversial. What makes it so are several thorny questions: How will the monitoring be structured? How will it be financed? Who will do the monitoring? This prompts a more general question. In view of all the effort, resources and controversy surrounding the creation and maintenance of a monitoring system, what might motivate an industry participant to take such a step? At least part of the answer is

23   This account is a modified and truncated version of Gunningham and Rees 1997.

that claims made by an enterprise may lack credibility. And from this credibility gap flows the need for some kind of independent confirmation of the industry's claims, by checking their accuracy, by monitoring the actual performance of partner enterprises, and so on. In other words, the environmental improvement targets set under the voluntary initiative may require the incorporation of a workable set of performance indicators. Again, these may take the form of quantifiable or qualitative measurements. In either case, it is arguable that they should be determined in advance of the scheme's operation, preferably in conjunction with the target-setting process.

But monitoring alone will not necessarily overcome the credibility gap, if the industry participant is still measuring its own performance. In many circumstances, but certainly not all,[24] independent verification will also be necessary. This is often painful. Opponents of verification highlight the risk independent audits pose to business autonomy and the confidentiality of trade secrets, as well as the danger that verification results could make them increasingly vulnerable to regulators, environmentalists and litigation. Yet, despite these and other concerns, the development of an independent verification capability is often of fundamental importance to the long-term viability of a voluntary initiative. Only then, for example, are community groups, NGOs, or even government agencies, likely to be convinced of the value of the arrangement. Suppliers and other commercial third parties will also want reassurance which can be provided, at least in part, by subjecting the measuring, monitoring and auditing arrangements to outside scrutiny. Certainly the verification process could be conducted in-house (for example by an 'arm's-length' audit team), but the closer the verifier is to the industry partner, the lower the credibility of its findings. Thus third-party audits provide far greater reassurance to outsiders than do internal audits.[25]

Environmental initiatives that include independent verification have a greater chance of success for two reasons. First, it builds in credibility and community and consumer confidence that the environmental claims are actually being delivered. This is important if industry intends to obtain a financial benefit from its environmental activities, even if this is not its primary motivation. For example, the consumers of environmentally preferred products require reassurance of the product's *bona fides*. Independent verification is far more likely to provide this than in-house verification. Second, knowing that the results of the environmental improvement activities will be periodically subject to external assessment provides an ongoing incentive for enterprises to deliver on their commitments (which brings us back to accountability). As we have seen, the problem with most independent verification initiatives, including AMICEM, is that they verify systems, not environmental outcomes.

Voluntary initiatives and environmental management systems

There is a striking similarity between the majority of the factors identified above, as key features of successful voluntary approaches, and the central ingredients of EMSs. As we

---

24 Some types of corporate misconduct are not amenable to external verification. Other tools need to be deployed (e.g. whistle-blowing facilities are considered to be important to some types of environmental misconduct).

25 However, even external audits may not be as independent as they purport to be. See Gunningham 1993.

have seen, such systems follow a defined sequence of steps which provide a structure for planning, implementing, reviewing and revising a system to address those parts of an enterprise's operations that can have an impact on the environment. In the case of ISO 14001, the further aim is to provide an international standard and a common (global) approach to environmental management and the measurement of environmental performance.

To meet the ISO 14001 standard, an enterprise must have a coherent framework for setting and reviewing environmental objectives, for assigning responsibility to achieve these objectives, and for regularly measuring progress towards them. It must also have appropriate management structures, employee training, and a system for responding to and correcting problems as they occur or are discovered. This implies documentation control, management system auditing, operational control, control of records, management policies, statistical techniques and corrective and preventative action.

However, while identifying environmental targets, performance monitoring, measuring and verification are all central to ISO 14001, third-party audits and transparency are not. These omissions have resulted in substantial criticism of the standard by NGOs and may well be addressed in the currently contemplated revisions of the standard. However, there is nothing in ISO 14001 that precludes greater transparency and third-party verification, and these elements can readily be incorporated by those who wish to do so. External pressures (e.g. public opinion, or pressure from trading partners), rather than ISO itself, will determine whether enterprises opt for such transparency of verification. If the experience of the quality standard series ISO 9000 is repeated, then supply chain pressure (as large companies, and multinationals in particular, require their suppliers to enter into contractual agreements committing themselves to become certified to the standard) may prove the most important determinant of companies seeking external certification, while NGO and community pressure may lead to greater transparency.

Finally, the most fundamental weakness of ISO 14001 as a stand-alone basis for a code of practice is that, as indicated above, it is a process standard not an outcome standard. This is not an argument against ISO 14001 per se but rather an argument for coupling ISO with agreed performance standards for the reasons set out in the section on environmental targets above. As Adams (1999: 116) put it:

> continuous measurable improvement in actual environmental outcomes is increasingly recognised as necessary to gain the trust of stakeholders . . . these efforts will be more successful the more the stakeholders are engaged in the process of setting, monitoring and continually improving the performance objectives. External verification is a crucial factor in making these voluntary efforts more credible and reliable.

This point is particularly apposite to AMICEM in its current form.

Do voluntary initiatives need to be combined with other policy instruments and external pressures?

Voluntary initiatives, such as the sorts of codes of practice contemplated by the mining industry, to the extent that they are viable, have the considerable advantages of providing greater flexibility to enterprises in their response, greater ownership of solutions that they are directly involved in creating, less resistance, greater legitimacy, greater speed of decision-making, sensitivity to market circumstances and lower costs. However, from a public policy perspective, such initiatives should only be preferred to the extent that they are demonstrably capable of delivering the identified environmental outcomes and achieving compliance on the part of target groups.[26]

As with other instruments, voluntary initiatives work better in some circumstances than in others, and not all industries lend themselves to such initiatives through industry associations. A review of the literature relating to voluntary initiatives (see e.g. Moffet and Bregha 1999) and industry self-regulation[27] suggests that necessary (but, as we will see, certainly not sufficient) conditions for the success of such initiatives are either: (1) a strong natural coincidence between the public and private interest in establishing such agreements, or (2) the existence of one or more external pressure sufficient to create such a coincidence of interest.

Circumstances where there is a natural and substantial coincidence between the private interests of individual enterprises and the public interest are often referred to as win–win. While such win–win opportunities do exist in some industry sectors and for some companies (Reinhardt 2000), they are often insufficient to prompt voluntary action, and are frequently overwhelmed by circumstances where no such self-interest exists.[28]

The second situation that is conducive to voluntary initiatives and self-regulation is where there are sufficient external pressures on enterprises or industry associations to give them an incentive to make such initiatives work. Those pressures might come from a variety of sources, and include the threat (actual or implied) of direct government intervention, broader concerns to maintain credibility and legitimacy (and, through this, commercial advantage), and the market itself. The likelihood of self-regulation and voluntary initiatives functioning successfully will necessarily vary with the strength of these pressures.

Probably the circumstances most conducive to industry self-regulation are those where an industry, or at least industry leaders, perceive the future prosperity and perhaps even the very survival of the industry as dependent on some form of self-control. Significantly, even where the industry is not directly connected with the consumer and is not purchasing directly from it, public pressure may still be crucially important provided the public concern is deeply felt. This is the situation in the case of the mining industry, where we have argued that the fear of losing the 'social licence to operate' is particularly strong.

Moving beyond the specifics of the mining industry, the effectiveness of external pressures brought to bear by consumers or the broader public will necessarily vary depending on the type of product, the type of market (e.g. the number of players, their size, import and domestic considerations, stability), the extent of public concern or 'outrage', and whether there is some natural affinity between consumer and industry

---

26 An OECD study shows that voluntary initiatives are often shaped directly by the policy environments from which they emerge. They tend to enhance the effectiveness of public enforcement, and enforcement strategy has shifted toward greater attention to and use of private compliance processes. See OECD 2001.

27 See Gunningham and Rees 1997 and references cited therein.

28 For example, in relation to forestry, while limited win–win options may exist, the fundamental fact is that clear-felling remains by far the most economic option, and it is highly unlikely that sufficient self-interest exists to replace it with less environmentally damaging forest practices.

interests (Webb 2002). Of course, where a combination of various external forces can be brought to bear, then the chances of successful self-regulation are likely to be higher than otherwise. Success is most likely where there are: a small number of firms in each sector; domination of each sector by large firms; sectoral associations able to negotiate on behalf of their members; and a sympathetic business culture. Beyond this, there are a range of problems common to many attempts to develop collective voluntary initiatives which will be further examined below.

Dealing with free-riders

As indicated above, the existence of external pressures (or a substantial coincidence between public and private interests in collective action) is a necessary but not sufficient condition for the success of voluntary initiatives and self-regulation. A range of other factors will also be crucial to the ultimate success of such initiatives.

Of these, perhaps the most crucial is the ability to stop free-riding. This is a form of collective action problem identified by Mancur Olson and others. The essential problem is that, although each individual enterprise may benefit from collective action if other enterprises also participate (as when all agree to participate in a voluntary initiative that will enhance the reputation and competitive position of the entire industry), each will benefit even if it does not participate, provided that others do.[29] It is rational, therefore, for individual enterprises to free-ride: to defect or engage only in token compliance, in effect seeking to benefit from the collective scheme without paying, or by imposing costs on others without compensation. The temptation may be particularly strong for firms with low public visibility and no brand capital, and where the costs of environmental improvement are substantial.

For present purposes, there are two main versions of the free-rider problem. In the first, all parties agree to the terms and conditions of self-regulation, but some merely feign compliance. For example, a self-regulatory programme addressed to environmental protection, such as the MCA's code of practice, confronts the problem that some participants will be tempted to take advantage of the willingness of other enterprises to spend on cleaning up the environment, while refraining from doing so themselves as a matter of rational, economic self-interest.

A second version of free-riding occurs where a part of the relevant industry simply refuses to sign on to the self-regulatory scheme. For example, 80% of the industry may agree to comply with a code of practice, but 20% may simply refuse to sign on and still get the benefits of collective action by their competitors. A failure to address the misconduct of the latter (which, since they are outside the code, is beyond the scope of the self-regulatory scheme) will almost certainly result in the failure of the code. This is

---

29   The logic underlying Olson's theory of collective action is identical to that of an n-person prisoner's dilemma (Hardin 1971). Note, however, that in a continuing series of two-player games, the best strategy is 'tit-for-tat': i.e. to co-operate in the first game, and to do whatever the other player did last time, from then on (Scholz 1984; Ayres and Braithwaite 1992). Responsible Care, in its present form (relying solely on moral succession without sanctions), lacks the characteristics of a continuing series game.

because those who sign the code cannot afford to be put at a competitive disadvantage against those who do not. Significantly, Esmeralda Exploration Ltd, the Australian mining company responsible for the disastrous Baia Mare cyanide disaster in Romania, was not a signatory to the relevant industry code and (even in the event credible sanctions existed) cannot be sanctioned under it.

In the first version, where all, or almost all, firms agree to participate, then in at least some circumstances free-riding may be capable of being contained through mechanisms such as peer-group pressure, shaming, or more formal sanctions. A crucial consideration will be the number of players involved. As Olson and others have pointed out, the greater the number of players, the greater the temptations and opportunities to cheat, and the greater the difficulties in identifying and controlling those who do. There is also greater difficulty in reaching and maintaining consensus with a large number of players.

Even where there are reasonably few players, success in curbing free-riding is by no means assured. Rather, it will be influenced by a number of other factors. For example, it has been argued that the ability to control free-riding increases as:

- Enterprises are aware of each others' behaviour and can detect non-compliance.

- They have a history of effective co-operative action (e.g. an existing association).

- Non-compliant behaviour can be punished.

- Consumers, customers or other clients value compliant behaviour and can identify compliant firms (with the result that free-riders can be controlled by markets, particularly where these are driven by consumer demand).[30]

While it would be premature to judge the success of AMICEM in terms of its success in avoiding free-riding, the experience with the closely related Responsible Care initiative is not encouraging. Recent studies of Responsible Care cited above suggest that free-riding is rampant. In the absence of mechanisms to prevent it, the risk of similar behaviour in relation to AMICEM is high.

In the second situation above, where a significant number of players refuse to join the self-regulatory programme, and cannot be induced to do so by threats or incentives provided by other players, then self-regulation in relation to programmes such as AMICEM and Responsible Care can only work if government intervenes directly to curb the activities of non-participants. While this may be viable in the circumstances of many developed countries, it is rarely a credible option in developing countries (see the following section).

Collective action problems such as free-riding are most likely to be overcome where the common interests of the group members are particularly strong (for example, where there is a 'community of shared fate' whereby the fortunes of any one company are tied to the fortunes of the industry as a whole, as described above).

---

30 Cohen (2002) points to the case of the Canadian Care Labelling programme, which has few free-riders, in part because of active lobbying by consumer groups. He also cites GAP Inc.'s Sourcing Principles and Guidelines and the Canadian Eco-Labelling programme as examples of codes that employ the market's power to curb free-riders.

Given the serious problems of free-riding, a prerequisite for successful collective voluntary initiatives will be effective monitoring and enforcement. Without it, free-rider problems in relation to self-regulatory codes such as that of the mining industry may be insurmountable, for reasons described above. We have seen that the range of enforcement mechanisms that might potentially be invoked is quite broad. At the lower levels it could include education, incentives (e.g. under Responsible Care, the sharing of information) and peer pressure (e.g. Responsible Care leadership committees). At the higher levels, sanctions might include removal of benefits (e.g. the right to use the industry logo), a requirement of public disclosure of breaches (making the perpetrator vulnerable to adverse publicity), or the taking of remedial measures (product recall, reparation of environmental damage). Breach of terms of a self-regulatory programme might also be construed as breach of contract, making a defecting enterprise liable in damages to the relevant self-regulatory body.

The ultimate sanction is often expulsion from the association, compliance being made a condition of membership. The impact of this will vary from case to case. Where an enterprise cannot effectively trade without industry membership, this sanction may be potent indeed, though in these circumstances serious concerns may be raised about restrictive trade practices and the contravention of any relevant anti-trust laws. Where expulsion will have little direct impact, associations will be reluctant to invoke it for fear of revealing their ultimate lack of regulatory clout. It is at this point that most collective voluntary initiatives are vulnerable to failure. If there is not the ultimate capacity to invoke sanctions at the tip of an 'enforcement pyramid' (Ayres and Braithwaite 1992; see Chapter 6), the credibility of sanctions at lower levels also weakens. This is a major reason why 'pure' self-regulation is rarely successful, and why there is a compelling need, even with many of the best of self-regulatory programmes, to complement self-regulation with some form of government or third-party involvement. We explore this issue further in the section below.

The role of the state and third parties

How can public policy best be designed, in order to take advantage of the strengths and virtues of collective voluntary approaches and industry self-regulation,[31] while compensating for its weaknesses as a stand-alone mechanism? This implies an underpinning of state intervention sufficient to ensure that it does operate in the public interest, that it is effective in achieving its purported social and economic goals and has credibility in the

---

31 Voluntarism and self-regulation are asserted to have a variety of benefits. Because standard-setting and identification of breaches are the responsibility of practitioners with detailed knowledge of the industry, this will arguably lead to more practicable standards, more effectively policed. There is also the potential for utilising peer pressure and for successfully internalising responsibility for compliance. Moreover, because self-regulation contemplates ethical standards of conduct that extend beyond the letter of the law, it may raise standards of behaviour significantly and lead to a greater integration of environmental issues into the management process. See Gunningham and Rees 1997; Lyon and Maxwell 1999; Moffet and Bregha 1999.

eyes of the public or its intended audience. As we have indicated, precisely what form of state intervention will provide the most appropriate underpinning, and indeed the extent to which such underpinning is necessary, is likely to vary with the particular circumstances of the case. Unfortunately, there are no magic bullets or universally appropriate prescriptions. However, it is at least possible to identify some of the most commonly important variables, and to illustrate by example how co-regulation might operate to optimal effect in particular circumstances.

In the case of codes of practice along the model of the minerals industry or Responsible Care, collective voluntary initiatives should ideally operate in the shadow of rules and sanctions provided by the general law, for these are the most obvious and visible (but not the only) means of giving regulatees the incentive to comply with the self-regulatory programme. Certainly, there is considerable evidence from a variety of jurisdictions that it is largely fear of government regulation that drives the large majority of self-regulatory initiatives, and it seems unlikely that they will perform well in the absence of continuing government oversight and the threat of direct intervention.[32] The virtue of voluntary codes in this context is 'their capacity to reach beyond government regulations and get industry to commit *of its own free will* to goals of improved environmental performance' (Wells 2001).

However, in the context of developing countries, the law is rarely a credible and effective policy tool, and environmental regulators are usually vastly under-resourced and sometimes vulnerable to capture and possibly corruption. Accordingly we must look elsewhere for means to bolster the effectiveness of voluntary initiatives.

Ideally, this might involve other forms of regulation. As Zarsky (1999: 49) has pointed out:

> without obviating the need for local regulation, there is a great need for an *overarching global framework* to define—and raise—the environmental responsibilities of foreign investors. Only by setting common responsibilities for all transnational investors will policymakers escape the competitive race for Foreign Direct Investment (FDI) which keeps environmental commitments 'stuck in the mud' [emphasis original].

However, it is also recognised that there has been little political will by governments for global and/or regional social regulation of investment, and for the moment one must look elsewhere for external controls on corporate behaviour.[33]

Conceivably, individual countries might choose to go it alone, and impose 'long-arm' legislation whose reach extends internationally to companies registered in that country. For example, Australia already has such statutes relating to sex tourism and bribery, and, as Bill Dee points out, 'the addition of environmental legislation applying to Australian mining companies operating abroad would not be unprecedented'.[34] A private members bill (the Corporate Code of Conduct Bill) is currently before the Australian parliament seeking to impose standards of conduct on Australian corporations undertaking busi-

---

32 See e.g. Davies and Mazurek 1996; Harrison 2001. Note also the evidence suggesting that domestic legislation is by far the most important influence on environmental management practices. See OECD 2000d.

33 In this regard extraterritorial liability on environmental issues in home-country courts might be an alternative way.

34 Personal communication, December 2000.

ness activities and employing more than 100 people in a foreign country. However, such legislation is unlikely to be implemented, not least because of the threat that any country doing so would be creating incentives for powerful corporations to relocate elsewhere, taking much of their capital with them.[35]

Sometimes, there may also be a possibility of harnessing third parties to act as surrogate regulators, monitoring or policing the code as a complement or alternative to government involvement. Indeed, it is arguable that self-regulation is rarely effective without such involvement. Thus Webb (1996), summarising the experience of the 1996 Canadian Symposium of Voluntary Codes, concludes:

> Meaningful involvement by consumer and other public interest groups is often what sets apart the successful codes from those which have received less support from government and the general public. At a time when citizens are better informed, more demanding and more sceptical of so-called 'elites' (government, industry, the academic and scientific communities etc) it is difficult to imagine a situation where a voluntary arrangement could succeed without meaningful community, consumer and/or other third party involvement.

The most obvious third parties with an interest in playing this role are sectoral interest groups such as consumers, trade unions or NGOs generally. This contribution may be through their direct involvement in administration of the code itself (in which case it has greater credibility as a genuinely self-regulatory scheme) or in their capacity as potential victims of code malpractice, to boycott firms that breach the self-regulatory programme. In the case of the mining industry, public pressure, fuelled by NGOs, and fear of losing the 'social licence to operate' are driving forces that provide incentives to the industry to develop and implement voluntary codes. Arguably, in the case of the mining industry the World Bank could also act as a surrogate regulator, insisting on certain minimum environmental requirements as a precondition for providing the financing that is essential for many projects in developing countries to progress.

Sometimes the role of third parties and government intervention can be combined. For example, as we saw in Chapter 6, the Indonesian government has had considerable success with its PROPER PROKASIH programme, in which companies are ranked according to their performance and the information made public.

Finally, the importance of utilising a broader regulatory mix cannot be overemphasised. Often, the best solution is to design complementary combinations using a number of different instruments. Thus self-regulation, government regulation and third-party oversight may be capable of being used in complementary combinations that work better than any one or even two of these instruments acting together.

For example, in the case of the chemical industry's Responsible Care programme, even though the industry as a whole has a self-interest in improving its environmental performance, collective action problems and the temptation to free-ride mean that self-regulation and its related codes of practice alone will be insufficient to achieve that goal. However, a tripartite approach involving co-regulation and a range of third-party oversight mechanisms may arguably be a viable option. This might involve: creating greater transparency (through a CRTK about chemical emissions), which in turn enables the community to act as a more effective countervailing force; greater accountability

---

35 However, such fears may be exaggerated, given that environmental costs are usually quite low, and other factors such as the fiscal regime are more likely to trigger relocation.

(through the introduction of independent third-party audits which identify whether code participants are living up to their commitments under the code, and which involve methodologies for checking and verifying that responsibilities are being met); and through an underpinning of government regulation which, in the case of companies that are part of the scheme, need only 'kick in' to the extent that the code itself is failing or when individual companies seek to defect from their obligations under it and free-ride.

Conclusion

To summarise, voluntary initiatives are unlikely to make a substantial contribution to improved corporate environmental performance as a stand-alone policy instrument. The evidence suggests that sole reliance on voluntary initiatives has generally proved insufficient to achieve an acceptable level of industry-wide compliance. For example, a KPMG Ethics survey of 1,000 corporations found that 58% of those who said they had a code did not have anyone designated to be responsible for ethics within the company. Again, a Canadian survey concluded that industrial sectors relying solely on self-monitoring or voluntary compliance had a compliance rate of 60% versus the 94% average compliance rating for industries subject to federal regulations combined with a consistent inspection programme (Pacific and Yukon Regional Office of Environment Canada 1998).

It is when such agreements are used in conjunction with state or third-party oversight, or as a form of co-regulation, that their prospects are somewhat more promising (Harrison 2001). Yet the retreat of the regulatory state has meant that, in many contexts, effective government oversight is lacking.

The problems of the retreat of the state are exacerbated in the case of foreign direct investment in developing countries, where the capacity of the state to curb corporate environmental excesses was never strong and is now being rapidly outpaced by the growth of FDI and the severity of environmental problems such countries face. In these circumstances, voluntary initiatives would seem to be a far less than ideal policy approach. There are very few credible alternatives, however. Certainly, we should not hold our breath waiting for effective environmental regulation in developing countries or for the establishment of a global regulatory regime.

In these circumstances, however imperfect, voluntary initiatives must be strengthened by coupling them with external pressures and oversight. This can take a number of different forms. In the case of the mining industry, the capacity for corporate shaming, the importance of reputation capital and protecting the social licence to operate (because what a company does in any location and with any stakeholder will contribute to the company's reputation worldwide) may be the points of vulnerability that give the necessary incentive to large mining companies to take a code of practice seriously and, in the longer term, to pressure their peers to take it equally seriously. In this exercise, governments can help either by direct regulation where this is practicable, or by initiatives that provide environmental information and rankings and which facilitate corporate shaming such as the PROPER PROKASIH programme in Indonesia.

Even here, those (often smaller) enterprises that do not have a corporate brand or reputation to protect, and those that operate from countries that do not take environmental responsibilities seriously and are insensitive to the particular means by which enterprises maximise their profits, will pose intractable problems. We are still far from having a viable solution to the environmental excesses of some components of corporate capital in a globalised economy.

COMMUNITY EMPOWERMENT AND REGULATORY FLEXIBILITY
EIPs and accredited licensing

In the early 1990s the Australian State of Victoria introduced a new policy instrument, the Environmental Improvement Plan (EIP), designed to reduce polluting emissions from major industrial sites. As described by the Victorian Environmental Protection Authority (VEPA) (1993b: 1), an EIP:

> is a public commitment by a company to enhance its environmental performance. The plan outlines areas of a company's operations to be improved and is usually negotiated in conjunction with the local community, local government, EPA and other relevant government authorities. Where possible, an EIP contains clear timelines for completion of improvements and details about on-going monitoring of the plan. Improvements may include new works or equipment, or changes in operating practices. Monitoring, assessments and audits are undertaken to plan and support these improvements.[1]

1   The large majority of EIPs include a brief description of the enterprise's operations; an overarching environmental policy or mission statement; a description of the relevant environmental legislation applying to that company and, possibly, specific licences and licence conditions; a description of the major (or perceived to be major) environmental issues or problems confronting the company; a series of environmental commitments or targets that the company intends to achieve including: performance targets (for example, a reduction in specific chemical pollutant over specific period of time); process targets (for example, a commitment to introduce particular environmental management strategies); or specification targets (for example, a commitment to introduce a particular pollution reduction technology); some form of mechanism for ongoing monitoring and reporting (to the Community Liaison Committee [CLC]) of company performance against their EIP environmental improvement targets; a proposed emergency response strategy (or strategies) for dealing with chemical accidents or spills; and details of the company's CLC, including membership and frequency and nature of consultations. Beyond this, the precise nature of an EIP is determined by the participants themselves: most importantly, the company, community representatives and VEPA officials, namely regional client managers and, on occasion, local government agencies. This extremely flexible approach enables EIPs to accommodate the wide variety of industrial and site circumstances of participants.

In 1994 VEPA went further, offering a flexible alternative to 'the standard prescriptive approach to works approval and licensing' (VEPA 1994: 1). This was made available to licensees that can demonstrate a high level of environmental performance and an ongoing capacity to maintain and improve that performance. Applicants for the new 'accredited licence' were required not only to implement an EIP but also to establish an environmental management system (EMS) and an environmental audit programme (EAP). Provided they are able to demonstrate to VEPA that they have done so to a sufficient standard, they are issued with a licence which grants them 'a high degree of operational freedom' (VEPA 1993a) and autonomy.

These two related initiatives represented significant departures from conventional command-and-control regulation in two key respects. First, they emphasised a systematic approach to pollution prevention, a form of process-based regulation[2] intended to influence management practices, to 'make industry think' and to encourage greater self-management. Second, they involved a significant shift from the traditional bipartite relationship between regulators and regulatees to a tripartite approach involving disclosure of information to, and consultation with, local communities. By doing so, VEPA moved substantially towards 'facilitative regulation' whereby stakeholders are involved in the development and implementation of co-operative strategies to improve environmental quality.[3] Victoria is not alone in progressing down this alternative regulatory path, and has since been followed by several other Australian jurisdictions.[4] But it was the pioneer, and remains the leading exponent of this approach in Australia.

In international terms, facilitative regulation in the form of EIPs and accredited licences is of particular interest for a variety of reasons. First, in contrast to the sorts of regulatory flexibility initiatives being adopted in the USA and Europe, the EIP is targeted not just at best performers who voluntarily aspire to beyond-compliance outcomes but also at environmental laggards who would not choose to adopt the requirements of this instrument voluntarily. As such, it provides perhaps the only available case study of the extent to which process-based regulatory flexibility can be extended beyond a select few, and used as an alternative regulatory strategy for mainstream, and even poor, performers. Second, both of these initiatives were introduced many years before the current rash of

2 Process-based standards specify the procedures to be followed in managing particular hazards, rather than the outcome to be achieved (though they can, of course, be combined with outcome-based standards).

3 This implies: 'identifying strategic alliances and forming or facilitating partnerships. Environment Improvement Plans . . . are key examples of the mainstreaming of this approach' (VEPA 1998).

4 In Queensland, South Australia and Tasmania, EIPs (or similar instruments) enable regulators to require various innovations, such as ongoing review, monitoring, efforts to bring about compliance and the reduction of environmental harm, or provision for the transition to an environmental standard. Relevant legislation includes: Environmental Protection Act 1994 (Qld) ss. 80–100; Environment Protection Act 1993 (SA) s. 54; Environmental Management and Pollution Control Act 1994 (Tas) ss. 37–39. See generally Bates 1995. The Queensland legislation requires that draft EMPs be made available for public consultation, and approval may be subject to appeal. Environmental Protection Act 1994 (Qld) ss. 85, 93; Environmental Management and Pollution Control Act 1993 (Tas) ss. 40–41. In Western Australia the 'Best Practice Environmental Licence' (Environment Protection Regulations reg. 51A made under the Environment Protection Act 1986) has many of the features of the Victorian accredited licence and some additional, novel characteristics, such as benchmarking.

regulatory flexibility initiatives in North America, most of which are still at their 'pilot' stage. As such, they are mature regulatory innovations from which much more can be learned empirically about the extent to which it is credible to shift from direct regulation to co-regulation and industry self-management than from the much more recent and small-scale experiments conducted principally in the USA. Third, the EIP experience in particular provides insights into the strengths and weaknesses of community-based approaches and the value of this form of informational regulation and regulatory pluralism.

The role of facilitative regulation

EIPs in Victoria evolved against a backdrop of profound change in the direction of environmental policy internationally, not least a growing disillusionment with command-and-control regulation and a search for more effective and less-cost alternatives. However, it was the history of community–company relations at the Altona chemical complex in Melbourne that provided the most immediate impetus for the introduction of EIPs. Wills (2000: 26) describes the role of the Altona chemical complex in shaping regulatory policy over several decades, particularly in relation to community consultation:

> Beginning in the 1960s, several adjacent chemical manufacturing plants were established in the western suburbs of Melbourne to use feedstock from the nearby Mobil Altona Refinery. Until the 1980s, local residents, although fearful of exposures to the hazardous chemicals used and stored at the site, and upset at periodic noise, odours and liquid discharges from the Complex, felt powerless to influence industry actions: the chemical industry was very important to the local economy, the command and control regulator, the EPA, was seen as remote and slow to respond to complaints, and there was no dialogue between industry managers and the community (Hardy, 1998). In the 1980s, a distrustful local community began to oppose all development proposals at the Complex. The stalemate was broken in 1989 by the formation of Altona Complex Neighbourhood Consultative Group (ACNCG), comprising industry site managers, residents and EPA and local government representatives.[5]

The Altona chemical complex experience was a powerful demonstration of the benefits of engaging a broader range of stakeholders in the regulatory process. It largely dissipated the conflict between industry and community and, in the words of one community participant, achieved 'a substantial reduction in emissions and other adverse effects from industry and a more responsible attitude' (Hardy 1998). It also created greater investment certainty for the chemical industry, as a more trusting local community ceased to oppose development proposals as a matter of course. A A$1.8 billion (Unglik 1996) investment at the complex shortly after the establishment of the new détente provides tangible evidence of this.

In introducing EIPs, VEPA sought to institutionalise the Altona experience by requiring participants to engage in a formal process of community consultation, in the expectation

---

5   See also Wills and Fritschy 2001.

that what had worked at Altona could be made to work elsewhere. This was to be achieved principally by requiring EIP participants to form Community Liaison Committees (CLCs) designed to bring the various stakeholders into an open forum to work through environmental concerns at a particular industrial site and reach agreement on key issues.[6]

According to VEPA (1993b: 1), the CLC:

> ideally would include residents from the local community surrounding the plant, key company representatives including senior staff, local government, EPA and other government agencies as appropriate. These key people would work together as a group . . . The CLC would develop clear objectives and operating procedures, and develop and monitor the EIP. The CLC would also provide a forum to discuss other issues of concern to its members. These may be outside the scope of the EIP, but still relevant to the local community and how it views the industry's operations. The CLC would also develop ways of keeping the wider community informed of its activities, such as the production of a newsletter.

The involvement of community groups in the regulatory process is a major departure from the traditional approach, which involves standards that are created, implemented and enforced by government. Under command-and-control there is little scope for community participation except through making complaints, or conceivably through a formal appeal process in relation to licence conditions. In contrast, the new tripartite EIP approach has several potential advantages.

First, community members are empowered to participate directly in dialogue and negotiations with companies about their environmental performance and how it might be improved. Second, in the process they gain access to new information about corporate environmental performance, much of it disclosed as a result of the measurement and self-monitoring requirements that are built into the EIP process. Thus CLCs may be viewed as a form of informational regulation and as a democratisation of the regulatory process, at least at a local level. Third, this dialogue facilitates greater understanding between industry and community (and sometimes local government)[7] about their respec-

---

6   These include negotiation, consideration and recognition of: company environmental performance, pollution incidents, their causes and corrective action taken; plant development proposals, pollution licence applications, training programmes and progress towards target levels of emissions; commercially confidential company information; a company's right ultimately to decide appropriate environmental management and protection measures, provided these are reported to the community; and (by VEPA) of company and community rights to detailed information about, and explanations of, its pollution control definition, monitoring and enforcement procedures.

7   In some EIPs, local government representatives have been very active participants in the CLC process. These have tended to be the larger, more prominent EIP sites, or those where there has been a long history of community agitation. For example, local councils are involved in the various Altona complex company EIPs and Ford's Geelong site EIP. The most common local government representative is the environmental officer or, in some cases, health and/or safety officials. Councillors themselves may also attend meetings, as is the case in Altona. In terms of the more high-profile EIPs in particular, council representation is a means of demonstrating to the broader community that their interests are being looked after by the council. This is in spite of the fact that, at least initially, some councils showed a degree of scepticism as to the merits of EIPs. A potential contributing factor in this regard may have been the fear of endangering a valuable source of rates.

tive goals. The presence of VEPA in those negotiations, acting as 'honest broker', capacity-builder and coach, facilitates co-operative engagement.[8] Fourth, from the perspective of VEPA, the involvement of community members provides an opportunity to create a third-party oversight mechanism as a complement to conventional regulation, and to leverage scarce regulatory resources.[9] Finally, CLCs provide VEPA with the opportunity to more effectively demonstrate and communicate its policies and programmes to local communities. The overall result should be improved corporate environmental performance, reduced conflict between the community and industry, and reduced costs to VEPA.

Through the EIP mechanism VEPA also sought to establish a form of process-based regulation whereby enterprises are encouraged to think through solutions to environmental problems in a new way, to systematically devise novel approaches consistent with risk management principles, and to focus their attention on important issues of environmental decision-making which otherwise might be ignored and which 'slip through the gaps' of traditional technology-based or performance-based standards. The premise is that those that engage in planning and certain processes will achieve better results than those that do not. The process-based components of EIPs (beyond community consultation) include the establishment of environmental improvement goals, the introduction of an ongoing monitoring programme, regular reporting of environmental performance, the development of emergency plans in the event of an environmental accident, and a commitment to assess new pollution control technology.

This process-based approach was taken substantially further with the introduction of accredited licensing which, in addition to the EIP, requires participants to establish an EMS (which may or may not be certified under ISO 14001 but which must be of a standard sufficient to satisfy VEPA guidelines). This EMS must itself must be certified and audited by a VEPA-appointed and -approved environmental auditor, more particularly one that has had no previous involvement in the design and implementation of the EMS in question.[10] Third-party oversight[11] is intended to provide an objective review of whether

---

8   The result may be to reach a mutually acceptable compromise and, in the words of VEPA, to 'reduce lengthy delays that sometimes occur when a community becomes concerned about industrial processes' (VEPA 1993b: 2).

9   In all these contexts, effective self-monitoring is particularly important because it enables community members, VEPA officials and corporate management to judge the extent to which a company is achieving mutually agreed pollution reduction targets.

10   The Environment Protection Act 1970 requires the applicant to have 'a suitable environmental management system in place which is certified by a person approved by the authority'. This is generally (but not exclusively) achieved by assessment against a recognised EMS standard such as ISO 14001. Such certification is carried out by an EPA-appointed Environmental Auditor (Industrial Facilities). Such auditors are people 'approved by the Authority. Certification must be conducted by an auditor who was not involved in the development and implementation of the EMS and who is free from bias and conflict of interest' (VEPA 1998: 2).

11   An innovative alternative that has been trialled in Alberta, Canada, is a peer evaluation system whereby each participating company agrees to receive the services of a certified independent auditor from a participating company in the same industry group. Whether such a system would work at least as well as one utilising auditors from outside the industry itself, whether it would result in collusion, or the converse (auditors from rival firms exploiting opportunities to disadvantage their rivals), it is too soon to say. This is an area where further empirical evidence is needed and where much may depend on the characteristics of the individual industry.

environmental requirements are being met and whether systems are being adhered to (Gunningham and Prest 1993). In essence, such audits entail 'the structured process of collecting independent information on the efficiency, effectiveness and reliability of the total . . . management system' (Health and Safety Executive, UK 1992: 65).

We identified the potential benefits of environmental management systems and their role in regulatory flexibility initiatives in Chapter 6. In particular, by creating a set of routines for collecting and evaluating information, and identifying environmental impacts, they not only 'make industry think' about its environmental challenges but also establish systemic ways of resolving them. As such, EMSs are potentially powerful management tools capable of helping an organisation achieve its environmental goals.

Finally, VEPA saw virtue in encouraging industry to take greater responsibility for its own environmental affairs, in facilitating greater industry self-management and internal responsibility (Victorian Legislative Council 1989: 1,537) and in encouraging beyond-compliance environmental improvement. Such an approach involves transferring greater responsibility for administering legislation and regulations from government to industry, and greater flexibility and autonomy for industry as to how it goes about discharging those responsibilities. Both EIPs and accredited licences facilitate this approach (and, in the case of the latter, also encourage continuous improvement and cultural change). Ideally, the result will be to ease the regulatory burden on VEPA, enabling it to refocus its scarce resources on other enterprises that posed a major environmental hazard but which were unsuited to the EIP or accredited licence processes.

Leaders and laggards

EIPs are targeted at three quite discrete groups. The first are environmental laggards that are conscripted to adopt an EIP. Statutory EIPs (or voluntary EIPs entered into as a result of VEPA 'arm-twisting', described below) are invoked to persuade particular poor performers to come into compliance. Indicators of poor performance are persistent regulatory breaches and, in some cases, past prosecutions; a high number of community complaints; and a prevailing management culture that is at best apathetic, and at worst antagonistic, towards the application of good environmental management practices. For enterprises with one or more of these characteristics, the short-term aim of the EIP is to minimise the risk of significant breaches of environmental law (that is, minimum levels of compliance), although in the longer term the aspiration may be continuous improvement and more ambitious environmental outcomes.

Laggards can be compelled to develop and implement an EIP by virtue of the statutory powers bestowed on VEPA by s. 3IC(6) of the Environment Protection Act 1970 (Vic).[12] A statutory EIP made under that provision includes requirements: that any relevant state environment protection policy, industrial waste management policy, regulations and

12 Under the Environment Protection Act 1970, s. 3IC, the Minister, on the recommendation of VEPA, can 'declare' an industry, giving firms within it a substantial incentive to develop and implement an EIP, because, should they refuse to do so, or fail to adhere to its provisions, they face a less palatable alternative: namely, to be made subject to the mandatory auditing requirements of s. 3IC(4).

licence conditions must be complied with; for the monitoring of compliance with the EIP; for the participation of the community in the evaluation of the performance in meeting objectives under the EIP; for the upgrading of plant and equipment to meet objectives under the EIP; for the assessment of new or emerging technology in the industry or in pollution control; and for contingency or emergency plans.

In practice, VEPA has found it largely unnecessary to invoke its statutory powers because it can almost invariably achieve the same end result[13] at less cost and inconvenience by various forms of arm-twisting. For example, it can invoke the threat of more stringent licence conditions, capable of being imposed by virtue of the very considerable discretion bestowed on VEPA client managers. A further threat available to VEPA is the potential application of a compulsory VEPA compliance audit,[14] which would not only be costly to the company but might form the basis for subsequent prosecution. The combination of these two potent and explicit threats (often reinforced by community demands for environmental improvements)[15] has been employed by VEPA and its staff to considerable effect. One corporate environmental manager described the pressure to adopt an EIP as follows:

> How it arose? We got browbeaten into it . . . A major issue was odour. We had various skirmishes with EPA. It was a confrontational period—but we took stock. We had a fairly good story, so there were two indignant sides and the lawyers were enjoying it really. Half the effort went on covering your backside—not really tackling the problem but just covering yourself legally. And the EPA was playing with community consultation—so we started to do it different.

VEPA officials estimate (there is no formal categorisation) that more than half of the total number of EIP companies fall into this category, and are in effect conscripts rather than volunteers.

The strategy of targeting laggards is a substantial departure from many of the regulatory flexibility initiatives that have been introduced over the past decade or so to overcome the alleged shortcomings of traditional command-and-control regulation. For example, the Clinton–Gore administration's 'Reinventing Environmental Regulation' initiative as exemplified by Project XL and the Environmental Leadership Program, has been confined to environmental best performers, on the assumption that this group need to be encouraged and facilitated to move further beyond compliance, while laggards should be subject to conventional regulatory strategies (see Chapter 6). That is, enterprises that choose to participate are rewarded with greater regulatory flexibility, but only those that are demonstrably already very good performers will qualify for entry. Part of the logic for this strategy has been that it is unproductive to *impose* a change in corporate culture and that only those that embrace the concept of regulatory flexibility willingly are likely to achieve genuine, long-term improvements in environmental management and

---

13 The exceptions being: (1) the single statutory EIP, which was only imposed as a last resort by VEPA after repeated environmental failures on the part of the company in question, and failures to respond to informal promptings to adopt an EIP; and (2) where the courts have required an EIP.

14 Environment Protection Act 1970, s. 31C.

15 Of the 30 or so conscripts, VEPA officials estimate that approximately ten experienced a considerable level of community agitation prior to the introduction of their EIP.

outcomes. Statutory EIPs clearly do not fit this particular policy mould, with implications that we will return to later.

The second group to which EIPs are targeted is genuine volunteers: that is, those that choose to adopt an EIP as a stand-alone innovation (as distinct from adopting it as one component of an accredited licence).[16] Participants in this second category are likely to be those with, at the very least, a good environmental performance record. From the perspective of VEPA, the aim of stand-alone voluntary EIPs is to facilitate and encourage good performers (those enterprises that are substantially in compliance, but not necessarily industry leaders) to adopt a more responsive and self-regulatory approach to meeting community expectations of improved environmental performance and, having done so, to go beyond compliance with existing regulatory requirements. The aspiration is to shift responsibility for future environmental improvements (at least in part) from regulators to industry, and to lock in a degree of community consultation and participation that had rarely been apparent in the past.

But since VEPA coercion or arm-twisting cannot appropriately be applied to volunteers that are already in compliance, other motivators are needed to induce their participation. Of these, by far the most powerful is community pressure. An antagonistic community can inflict considerable damage on a company, particularly if it has a prominent image or brand name to protect. Company management may perceive that its 'social licence' is under threat, prejudicing its public and commercial profile, making it more difficult to obtain planning approvals, and more vulnerable to the imposition of onerous regulatory conditions. One environmental manager put it more graphically: 'The community expectation is river quality will improve and there will be a slow decline in the level of emissions—*so if we asked to put more in the river it would unleash a holocaust*' (emphasis added). Thus, in this context, company management may view the EIP process as a means of ensuring the future social and regulatory viability of their operations. As one respondent pointed out:

> Why did we get involved in EIPs? That's easy—to provide certainty to the planning approval process. We had already had one surfactant plant knocked by community opposition. We saw EIPs as an opportunity to work proactively with the community. It would allow us to proceed [with our developments] without excessive delays.

The third group, accredited licensees, are also volunteers rather than conscripts. For them, an EIP is one of the three preconditions for obtaining this form of licence, along with an environmental management system and an environmental audit. Applicants for an accredited licence must also demonstrate a high level of environmental performance and an ongoing capacity to maintain and improve that performance. Those that can satisfy these criteria are a highly select group. They are, as one VEPA official put it, 'the best of the best', and being awarded an accredited licence is formal recognition of this fact.

However, obtaining an accredited licence can involve considerable costs for enterprises, in preparing the necessary documentation, negotiating with VEPA, and in demonstrably satisfying the three formal criteria for obtaining this form of licence. As with

---

16   VEPA officials estimate that, of the total number of EIPs (as we noted earlier, currently 56), approximately ten fit under the definition of stand-alone volunteers. There are also 17 accredited licensees.

stand-alone voluntary EIPs, some additional incentives will be necessary to persuade firms to participate. But, self-evidently, mitigating community pressure will likely be insufficient, since a stand-alone EIP will serve this purpose without the need to adopt the additional requirements of an EMS and environmental auditing. Among the additional inducements to take up this form of regulatory flexibility are licence fee reductions, less inspectoral scrutiny, the capacity to circumvent traditional (and potentially tortuous) works approval processes, the simplification of the actual licence conditions, greater flexibility and autonomy through options such as 'bubble' licences, and the prestige, external recognition and environmental credibility associated with obtaining an accredited licence.

Findings

Evaluating the effectiveness of EIPs and accredited licensing is a complex exercise, for a variety of reasons. First, they can be tailored to a variety of industrial circumstances, so one is not really comparing like with like. Second, VEPA itself has not conducted any formal evaluation of these instruments, nor does it keep any statistics that could be used to evaluate their success. Third, different CLCs may have different environmental concerns and therefore seek to impose different environmental management priorities, making it difficult to identify appropriate success measures.

Given the serious methodological challenges, our best (albeit far from ideal) option in evaluating the success or otherwise of the EIP and accredited licence programmes was to interview a representative sample of the stakeholders involved, especially industry, VEPA and community representatives, using a common set of evaluative questions. The responses to these questions (checked where possible against alternative sources of information, and sensitive to the potential bias of some stakeholders)[17] form the basis of our discussion below.[18] As we will see, EIPs had a different impact on conscripts and volunteers, and accredited licences were beset by a number of shortcomings quite specific to this form of regulatory flexibility. For these reasons, we maintain the threefold classification developed in above.

---

17   We were conscious that some stakeholders (certainly companies and, arguably, VEPA) might have had a vested interest in suggesting that these mechanisms had been successful. However, this was not the case with other stakeholders, such as community groups, suggesting that this response was a genuine reflection of respondents' perceptions. Nevertheless, the support of some groups was much more qualified than that of others and, as we will see below, the discernible differences between groups may in itself provide insights into both the strengths and limitations of EIPs.

18   This evaluation is based in part on interviews by the authors with relevant stakeholders, including 33 enterprise representatives (of which nine were accredited licensees), 13 VEPA officials, five community representatives and three local government officials. Community representatives were deliberately under-represented in our sample, since we chose to rely on the previous work of Wills (2000) on CLCs, focusing our own fieldwork on the other aspects of EIPs that had not previously been investigated.

Evaluating EIPs

In evaluating the success of EIPs, there was a marked difference of opinion between VEPA's centralised policy officials and regional field officers. The former, while acknowledging that EIPs are but one of a variety of possible policy responses, nevertheless took considerable pride in them. For example, one respondent claimed that:

> EIPs are by far and away the single most effective policy instrument we have introduced over the last ten years. Yes they have their limitations, but no other policy has the capacity to transform company environmental performance the way the EIPs do.

However, VEPA client managers (who, it must be said, have much greater day-to-day practical experience of corporate behaviour) were more circumspect in their evaluation and made an important distinction between the impact of EIPs on leaders and laggards, or, to put it differently, on volunteers and conscripts.

In the case of conscripts, most VEPA client managers expressed scepticism about the ability of EIPs to bring about fundamental changes in company environmental attitudes or behaviour. A substantial number considered that the impact of EIPs was mainly at the margins. As one put it: 'the EIP takes a previously antagonistic relationship . . . and you start to get positive involvement. But it's not always possible to get dramatic change—you need to begin with small steps, to keep them on-side.' The common view was that companies that were pressured into adopting an EIP because of their poor environmental track record were not somehow redeemed by the process, and for this reason still required a similar level of inspectoral supervision. For example, VEPA client managers pointed out that conscripts 'don't have a lot of commitment to environmental performance' and that 'it's hard to be confident about their engagement . . . we have to keep a close eye on them'.

Similar views were held by many conscript enterprises themselves. Most showed no inclination to shift voluntarily beyond compliance and a number conceded the need for external VEPA scrutiny to keep them from defaulting.[19] For example, one industry respondent acknowledged that 'they've got to keep us on our toes', the implication being that, if they did not, default would be common, while another provided a more concrete illustration: 'storm-water run-off wasn't well addressed. [The] community [was] not interested, and didn't pick it up. It was like an admission of failure. The EPA made it a condition of our licence.' And, because there was a reluctance to embrace environmental improvements voluntarily, such improvements as were achieved by conscripts rarely went much beyond compliance. There are a number of EIPs, for example, that have not progressed substantially beyond the original, relatively limited objectives. Several VEPA client managers highlighted examples where, once very modest environmental improvements had been achieved, further progress had effectively stalled. For example, one noted that a particular enterprise 'haven't grasped the concept of continuous improvement'. This was not assisted by a waning of community engagement (a matter to which we return below).

---

19 In a few cases, conscripts had gone further and progressively expanded the scope of their EIP to embrace a much more ambitious range of environmental improvement targets. Even here, however, there was a reluctance to abandon totally the regulatory safety net.

Community representatives themselves had similar reservations. While generally posi-tive about the achievements of EIPs, some had the distinct impression that the enterprise involved was 'talking down to them' at meetings. Some also expressed concern that com-munity meetings did not provide them with a genuine opportunity to influence the direction of EIP policy, such as improving targets and/or management practices, and many were left doubting the level of commitment of enterprise management to signifi-cant environmental improvements. One community representative voiced a common concern as follows: 'I don't think they have a lot of commitment for environmental performance. I suspect they were pushed into the EIP process.'

But, even if they were not capable of embedding a greater environmental awareness within corporate culture or in ensuring beyond-compliance improvements, EIPs still had two important virtues. First, they created a greater awareness among participants of their environmental obligations and provided a means by which client managers could focus their industry clients on specific measures that were necessary to bring them up to the minimum legal standard. Second, they empowered local communities to bring more effective pressure on those companies, reinforcing and complementing the inspection activities of VEPA. Thus an EIP may help to bring a recalcitrant or incompetent enterprise into compliance, albeit only when VEPA officials *and* the community provide monitoring and enforcement support and help to prevent backsliding. One VEPA regulator summed it up as follows:

> I would say that EIPs make my job easier. I still have to front up, there's still the threat of sanctions, but it gives me something tangible to aim for . . . The EIP is checked out when there is a significant complaints issue. If worst comes to worst, I can always modify the licence conditions to [reflect] the EIP . . . At the end of the day [however], EIPs don't work without community input.

In this context, community and VEPA pressure were commonly perceived to be mutu-ally reinforcing. Indeed, community representatives almost unanimously regarded VEPA pressure as essential to complement their own role, particularly to push the EIP process along to deal with target-setting and new environmental issues which many community representatives felt inadequate to deal with.

Even with such pressure, the outcome of the EIP process was not guaranteed. Much also depended on the views of individual decision-makers, and on how tensions between centralised officials and local management were resolved, as this interview with a plant environmental manager illustrates:

> It was prickly for a while, and awkward. Both sides knew it was not working. The directors were conscious of the due diligence clauses in regulation. I had the job of talking to the community committee about what we were going to do to improve—but EPA is sitting in . . . reserving the right to use it against you, and the company lawyers are saying 'admit nothing'. It was a no-win situation by being cautious so I went the whole hog—and the EPA held off—we told the community everything—all our dirty washing; over the course of three or more years the community became very supportive. I did get support from the general manager to go the whole hog but head office had distinct discomfort because of a lack of familiarity with local concepts.

In the case of volunteers who adopt a stand-alone EIP, the experience of VEPA client managers was much more positive, and many of them pointed out that it was enterprises that had voluntarily sought EIPs that had generated the most impressive environmental outcomes. The difference in attitude and performance of the latter group is reflected in the following description by a VEPA client manager of the impact of an EIP on one enterprise:

> [The company] has voluntarily embraced the community, and it has changed their culture. I think a big part of this is that senior management attend the meetings . . . They are exceeding expectations. Now that odour and noise are out of the way, the CLC has turned its attention to other issues. It's been a two-way process—as the community members have been educated, and become more confident, they have sought to tackle more complex [environmental] issues . . . I would like to take more of back-seat role, particularly if they start using audits.

Most enterprises adopting voluntary stand-alone EIPs also regarded them as an effective environmental policy instrument. While one should retain a healthy scepticism for potentially self-serving success stories, most enterprise respondents were able to provide detailed accounts of their environmental achievements subsequent to adopting an EIP. In some cases they were able to cite firm figures (a 62% reduction in waste-water, a halving in the opacity of fumes over a three-year period, major land rehabilitation, etc.) as well as less tangible but no less important indicators of success. For example, one industry respondent of an Australian subsidiary of a major international enterprise described the improvements achieved as follows:

> The EIP has transformed our operations. We are far and away ahead of our global colleagues. When I visit our company's other international sites, they are amazed at what we have done here with our EIP. They're still living in the dark ages—the thought of opening a dialogue with community groups is not even on their agenda. Even in Europe, they admire what we have achieved— this is a regular topic of conversation at international meetings . . . not that they are planning to follow our lead any time soon, but, still, it has an impact on their thinking.

In other cases, they pointed to the commitments they had made in their EIP which went very substantially beyond their regulatory obligations and commonly embraced issues not included in law at all (for example, to contain 92% of sewerage spills within 4.25 hours of receiving the report, to have less than 2.2% of trade waste customers recorded as being non-compliant, to reduce solvent usage in the painting process by 3% each year, to reduce the amount of prescribed waste generated on the site by 5% per year, etc.). Although we were not able to obtain any reliable and independent hard data on this issue, our interviews with a range of different stakeholders did suggest that such performance targets were largely (although certainly not entirely) achieved in the long term, with the enterprise facing an uphill battle explaining to an aggrieved local community any substantial departure from them.

It was largely for this reason that the majority of community participants in CLCs were broadly supportive of EIPs. One community representative told us the EIP process 'had greatly exceeded our expectations'. Another reported that:

> In the first couple of years the air was noticeably cleaner. People would come up to me in the street and thank me for the work we [the CLC] had done . . . Mostly, they commented on the absence of smell and noise. But there was also

an improvement in the general amenity of the complex—there was less rubbish lying around, and the whole site looked neater.

However some expressed concern that the rate of improvement gradually declined after the most obvious problems had been addressed, possibly because the early improvements are relatively easy and cheap to achieve, but once the low-hanging fruit has been picked further improvement becomes increasingly expensive. A declining rate of continuous environmental improvement may also be exacerbated by waning community interest. One community respondent described this phenomenon as follows:

> I would like to see tougher targets embodied in EIPs. We got a good reduction straight away. But it seems to take a lot longer for things to improve from there . . . it's like the 80/20 rule—they're prepared for the easy gains, but don't want to put in that last 20%.

A consistent picture emerges as to why EIPs were substantially successful in bringing about improved environmental performance beyond that attributable to 'business as usual', with only a few differences of emphasis between motivators of leaders and laggards.[20] For many, the strongest motivator for improved environmental performance was community involvement, which often changed the power balance in the enterprise, giving greater influence to environmental managers, and an impetus for staff to stay focused on their EIP commitments. For example, one manager pointed out how: '[the EIP] has taken on a life of its own—there is now an internal compulsion. What has been the big change? A commitment to the community.' Others emphasised how community consultation 'affects what we prioritise and how we approach projects', and provided graphic and detailed illustrations of how this had occurred. In a similar vein, one environmental manager contrasted decision-making processes in his company before and after the introduction of an EIP, pointing out how:

> In the old times, we scoped the project, put it together, got Board approval, did the costings, brought it back to the Board for final costings—*then* announced what we were going to do. And then maybe the regulators had an inkling, certainly the community didn't. Now, once we have done the preliminary scoping and got the preliminary OK, we brief the community committee, and ask their issues and concerns, then take these on board. They wanted no more effluent so the challenge was, can we do this and still build the new [installation]? And, because we knew this early, we could go to the designers and ask ways to achieve minimum water usage and how to offset this elsewhere in the plant. We achieved this, got environmental approvals from the regulators and had no objections from the community [emphasis original].

Company middle managers, in particular, highlighted the leverage that the EIP had given them and how it had enabled them to focus higher management's attention on the environmental improvement process. As one described it: 'for the first time, we were sitting around the table with senior management. They are engaged in the process now . . . This has made my job a lot easier.' Another stated that:

---

20 Of course, it is possible that improvements might be attributable not to the adoption of an EIP but to participation in other initiatives such as the chemical industry's Responsible Care programme. However, the vast majority of company respondents asserted that it was indeed the adoption of an EIP that had led to such improvements, and gave a number of reasons, described in the text below, why this was the case.

[The CLC] most definitely has senior management involvement. We provide the support, and a management representative and the environmental officer attend. This has been a big change . . . management was not very involved with ISO 14001 [their EMS]. But in the community meetings, they have to deal with the issues . . . This makes my job [as environmental officer] easier—it gives me more authority . . . The other plus is that management has to commit itself to meeting resolutions. This makes them more interested in tracking our environmental performance—their reputations are on the line.

The significance of securing senior management commitment cannot be overemphasised, as almost all studies of environmental decision-making confirm. Without such commitment and leadership from the top, many environmental management initiatives founder and fail. Yet getting senior management attention is not easy, and commonly environmental issues do not even appear on the radar screen of senior decision-makers. Certainly, the EIP did not succeed in generating such attention in every case and we noted earlier how there is sometimes a large gap between management attitudes at local level (where management are likely to be particularly sensitive to community pressure) and at head office. Nevertheless, particularly where senior management were present at CLC meetings (and so encountered at close quarters the concerns of the community) and where they personally signed off on the EIP (and so had direct obligations under it), the EIP was likely to have a substantial influence on them.

Three other aspects of the EIP were also influential on corporate behaviour. First, as a form of process-based regulation, EIPs frequently generated greater environmental commitment within the enterprise. Certainly for volunteers, and possibly for conscripts that were incompetent rather than recalcitrant, the very process of engaging in environmental planning, identifying environmental impacts, or in designing an emergency response plan in and of itself, had the capacity to generate change. This came about most commonly by transforming a previously isolated, haphazard approach to environmental issues into a systematic one which, because it was initiated largely from within the enterprise, led to greater management 'buy-in'. As one company representative put it: 'We always had a strong environmental ethic . . . What the EIP did was formally channel it better—it led to better improvements. For our management, it is a non-prescriptive way of tackling things.' Another pointed to how the EIP 'has had a number of impacts: greater environmental improvements; our relationship with the community; competition with others to be seen as the best corporate citizen. Some of these things may have happened in any case, but the EIP is better than an ad hoc process.' And a third, in acknowledging their prior marginal environmental track record, recounted how the EIP had 'focused the attention of senior management on environmental issues' and that this had 'brought about real changes in the way we do things—we try to anticipate problems more'.

Second, over and above such process-based changes (and in contrast to initiatives such as ISO 14001 and Responsible Care), the EIP also requires identified environmental outcomes. That is, enterprises committing to an EIP must meet specified performance targets within a specified time-period (for example, they may commit to upgrade equipment to meet objectives under the plan, or to meet specified emission or waste reduction targets). And, once those targets have been agreed and incorporated in the EIP, there are a number of compelling reasons why they are likely to be complied with. Not least the fact that these commitments are given publicly, and represent the negotiated outcomes of agreements made at the CLC with the community and VEPA, places considerable pressure on the enterprise to deliver. This, in conjunction with the obligation to self-monitor regularly and report to the CLC on progress,[21] serves as an important accountability and shaming mechanism, and as a vehicle for negotiation about what remedial action should be taken. One industry respondent stated that 'we wanted to set quite ambitious targets . . . [for example] a 64% reduction in waste-water . . . Once these were out there, we had an obligation to deliver on our promises.'

A third driver, largely confined to potential laggards, was ongoing VEPA inspections. As one industry respondent noted, 'our EIP targets are incorporated into our licence conditions—the EPA can check them any time'. Indeed, it was a VEPA official who provided a compelling argument for this form of EIP compliance:

> How do we know for sure [they are complying]? First, EPA site inspections address EIP obligations in conjunction with statutory obligations. Second, we build reporting mechanisms into the EIP targets—again this is something that is checked by inspectors. Third, we invite critical feedback from the Community Liaison Committee members—they can ask our inspectors to investigate any possible breaches.

To summarise, while the success of EIPs was more pronounced in the case of volunteers than of conscripts, even in the case of the latter EIPs helped to bring about a significant improvement in environmental performance. The main distinctions between the two groups were twofold. First, EIPs did much to bring conscripts up to compliance, but not much beyond, whereas in the case of volunteers they commonly achieved much more far-reaching change, and beyond-compliance outcomes. Second, while EIPs often succeeded in embedding environmental considerations in the corporate culture of volunteers, there was much less internal change in the case of conscripts. In the case of the latter, without continuing VEPA oversight and pressure, there was a substantial risk of backsliding.

Evaluating accredited licences

There are many purported benefits associated with regulatory flexibility initiatives such as accredited licences. These include: greater industry autonomy in how to achieve agreed environmental outcomes; greater transparency; fostering a climate of co-operation between the licensee and the regulator; enhanced credibility with the local community; fee reductions; simplified licence terms; and expedited works approval processing (DEP 1998: 9). Proponents also hope that the greater flexibility provided by such initiatives, in conjunction with the more systematic approach to preventative issues resulting from the introduction of an environmental management system, will generate not only better environmental outcomes but also continuous environmental improvement.

However, our research suggests that many of these benefits are more apparent than real. The most striking limitation of accredited licences is that very few enterprises have

---

21 While the accuracy of such internal enterprise compliance reports might be questioned, in practice there were strong disincentives to falsification. There was broad agreement that 'trust' between enterprise and community was a key feature of the EIP process, and that, should this be broken (for example by falsification of enterprise monitoring), then the corporation's credibility and reputation would be threatened, not just with the community but also with VEPA, and perhaps even with shareholders.

adopted them. In Victoria, after nearly eight years, only 17 companies (out of some 1,200 'scheduled premises') have enrolled in the VEPA's accredited licence programme (with a further six in preparation). After almost three years of operation, only one company has taken up a best-practice environmental licence under the comparable Western Australia scheme (although two others are in the pipeline). This is consistent with the similarly disappointing participation rate in most of the American regulatory flexibility initiatives, such as Project XL, described in Chapter 6. The explanation lies largely in the inadequacy of the incentives for participation. Adopting the three cornerstones of accredited licensing can be expensive, and there are additional transaction costs in negotiating the terms of a new licence with VEPA, raising the question of whether the benefits of so doing outweigh the costs.

Based on our fieldwork, the various inducements for leading enterprises to take up an accredited licence are insufficient to overcome the costs of doing so, whether used alone or in combination. For example, none of our industry respondents identified licence fee reductions as a significant motivator for participation, the amount saved commonly being described as 'fairly minimal'. Nor was the promise of less inspectoral scrutiny viewed as important, because most potential accredited licensees (which by definition all have a strong environmental track record) did not regard VEPA inspections as a threat.[22] And while proponents of accredited licences hoped that participating companies would have greater flexibility to generate cost savings through cleaner production and pollution prevention initiatives, this was not the view taken by companies themselves.[23] A much greater incentive was the capacity for accredited licensees to circumvent traditional (and potentially tortuous) works approval processes, which require a substantial investment of time and resources even before the actual construction begins, and are a source of considerable uncertainty and aggravation for many companies.[24] A similar benefit was a simplification of the licence conditions, with accredited licences being a fraction of the size of traditional ones, and much less prescriptive.[25] A further, albeit less tangible, benefit was the degree of prestige, external recognition and environmental credibility bestowed on participating companies.

However, even the strongest of these incentives rarely seemed sufficient to induce companies to seek an accredited licence. Rather, they persuade enterprises that have already introduced the main component of accredited licensing for entirely other reasons to seek the additional gains consequent on obtaining such a licence, almost as an afterthought. That is, many accredited licence participants had initially responded to

22 By way of example, while a pulp mill in Washington State, USA, would anticipate some sort of inspection every 6–8 weeks and a 'serious' inspection (for example, one lasting three days and carried out by a team of inspectors) each year, such a mill in Victoria would anticipate a single four-hour visit by an inspector (who will probably lack any expertise in that industry) once a year.

23 On the contrary, the large majority of industry respondents considered that substantial environmental improvements did not come cheaply, and required a significant injection of capital and resources. The empirical evidence on this issue is not consistent, with Gouldson and Murphy (1998: 85) suggesting that firms do find process-based cost savings in many circumstances.

24 Greater planning certainty for works approval therefore not only provides a short-term reduction in approval costs but also the opportunity to exploit the productivity benefits of initiating the proposed changes sooner.

25 The granting of bubble licences, facilitating 'whole of plant' regulation rather than specific

external pressures to develop an EMS (most commonly international supply chain pressure, or the broader reputational assurance concerns of their parent companies). It was only subsequently that their attention was drawn to the potential advantages of accredited licences. A typical industry response was the following: 'the EMS came first. This was a directive from our parent company in the USA. Then came the accredited licence.' Similarly, another respondent described how: 'Our plant had already achieved ISO 14001 accreditation. This decision wasn't taken in response to local regulations—it was purely a response to international developments. It was only subsequent to this that we contemplated participation in Best Practice Environmental Licensing.' And a third made the same point as follows: 'When we put in place our EMS, we hadn't even heard of accredited licences. It wasn't until later that we saw the potential advantages. Yes, these could be attractive . . . but, in our case, we haven't modified our [environmental management] strategies. It's been a case of "keep up the good work".'

Perhaps surprisingly, most accredited licensees had not previously developed the second cornerstone of accredited licensing, an EIP. One might have anticipated that environmentally proactive enterprises would first take on the basic process-based initiatives required by an EIP, and that subsequently the more advanced among them would adopt a formal environmental management system. However, this was certainly not the norm,[26] and for the most part an EIP was developed specifically in order to obtain the accredited licence. This might have been regarded as a positive in public policy terms because an EIP, as we saw earlier, can make a substantial difference to a firm's environmental performance. And, as a public commitment to specific environmental targets, it offers something crucial that an EMS generally lacks: performance standards rather than merely process-based standards.

In practice, however, most industry respondents considered that preparation of public commitments was more significant in 'giving voice' to their intended environmental improvements than in creating new and more ambitious environmental targets. As one industry respondent explained: 'our EMS already had [environmental improvement] targets built in. Sure, we might have tweaked them a bit, but they were basically the same.' Moreover, unlike most EIP participants, accredited licensees, being among the top environmental performers, were not generally subject to community pressure either through the CLC or otherwise. In fact, their biggest challenge under the EIP was often in actually *creating* (and then maintaining) a CLC, rather than meeting its demands. Indeed, recognising this difficulty, EPA officials have adopted an extremely broad and flexible definition as to what actually constitutes a CLC in this context. As one EPA official described it: 'every single accredited licence has some form of a CLC, no matter how informal. Admittedly, some of these are pretty marginal. But it still counts.' Thus the very foundation of the EIP process can become tokenistic in the context of accredited licensing.

Nor did the third cornerstone of accredited licensing, the requirement for an EAP, add anything substantial. Specifically, an accredited licensee is required to have an EAP which, while allowing for individual variations and circumstances, will be assessed by VEPA in

26 Of the 17 current accredited licence recipients, relatively few (less than one-third) received their licence *subsequent* to the adoption of an EIP. Similarly, few of the 55 or so EIP participants have gone on to achieve accredited licence status. The exceptions tended to be high-profile companies in environmentally sensitive areas such as the chemical industry (for example, Kemcor, the first accredited licensee, put in place an environment improvement plan three years before it obtained an accredited licence).

terms of how it addresses: (1) EMSs (applies to all applications); (2) compliance with existing regulatory obligations (applies to all applicants); (3) extent and sources of waste (applies to most applicants); (4) adequacy of risk management (applies to most applicants); and (5) the extent of environmental impact (applies only to applicants with the potential for large environmental impacts) (VEPA 1998). However, this adds little to what an accredited licensee would be obliged to do already,[27] and for this reason it is unsurprising that no industry respondent highlighted its EAP actions beyond that which was required as part of an EMS.

In one respect, it is possible that the Western Australian best-practice licence programme may push participants further than its Victorian counterpart, because the former also requires participants to undertake a range of activities beyond simply adopting an EMS, an EIP and audit. These additional processes in themselves might encourage improved environmental performance.[28] Of these, probably the most important is benchmarking against industry best practice. However, what impact this has in practice is not known. With only two participants currently in the Western Australian programme, and the most advanced of these only now in the process of taking stock of what impact the programme has had, we can only speculate as to the value of industry benchmarking. We note, however, that in other contexts it has proved an important motivator and tool for improved environmental performance for companies that aspire to be environmental leaders.

To summarise, the various incentives offered to environmental leaders to become accredited or best-practice licensees rarely proved sufficient to induce them to do so. Rather, the large majority of such licensees had enrolled only after implementing an EMS for other reasons, such as supply chain pressure or the dictates of head office. Taking up an accredited licence also required them to develop an EIP, but for the most part this did not cause them to change their environmental targets or how they achieved them. Nor did the third cornerstone of accredited licensing, the independent audit, add anything substantially new. Adopting an accredited licence did provide certain advantages to participating companies, particularly in terms of avoiding the works approvals process, obtaining simplified and more flexible licence terms, and obtaining external recognition for their environmental achievements. As such it provided welcome resource savings and public relations benefits. Viewed in these terms, accredited and best-practice licences are more a reward than an incentive. As such, they do not provide the sorts of environmental improvements, continuous improvement or cultural change that proponents of regulatory flexibility initiatives generally regard as the central contribution of this approach.

---

27 Accredited licences already stipulate that EMSs must be certified by an independent auditor. A component of ISO 14001 is that the relevant enterprise must be in compliance with all existing legislative obligations. Similarly, an assessment of waste, risk and environmental impact are also key components of EMSs.

28 In Western Australia, components of the best-practice licence include: an environmental policy; clearly defined environmental performance objectives; an environmental management manual; an environmental audit plan; an EIP; an environmental responsibility chart; and a system of control and verification of environmental actions.

The benefits and limits of community participation

In the case of both leaders and laggards, the EIP process was successful in generating dialogue and better understanding between enterprises and their local communities. Most community members interviewed considered the EIP process to be a valuable and worthwhile exercise from this perspective (as well as leading to real improvements in environmental outcomes for participating companies), as did company representatives. Even where there were substantial differences between the two sides at the beginning, constructive outcomes were commonly achieved, as the following anecdote illustrates:

> When we first met with the company, there was a lot tension. One manager in particular was very aggressive, and kept on interrupting and raising his voice. It seemed like we had nothing in common . . . we became disillusioned with the whole thing. Then, at the next meeting it looked like it would collapse— we [the community representatives] were ready to walk. And yet it was this manager, the same one, who suggested a way forward . . . As it turned out, he became our most enthusiastic supporter. We haven't looked back since then.

This is consistent with the findings of Wills's (2000) much more detailed study of CLCs. Almost all (32 of Wills's 35 interviewees) reported gains from the consultation process, usually in the form of better communication and relationships between the three parties, and more operational feedback to the firm managers concerned.[29] Questioned about information exchanges, survey respondents reported that community consultation resulted in the exchange of much additional information about industrial operations, pollution impacts, pollution control and each other's perceptions and actions. Industry and VEPA participants reported that they were better informed about community attitudes to, and fears of, industrial sites. Community participants reported learning most about plant operations and firms' behaviour; but, while many found their industry counterparts frank and reasonable, a considerable number remained suspicious of enterprises' motives. The survey results suggest that the consultation process commonly leads both community and industry participants to greater acceptance of the other side's situation and concerns.[30]

Yet the very process of generating, and subsequently maintaining, a substantive degree of community engagement is problematic, as both enterprise representatives and VEPA client managers pointed out. In some circumstances, no local community exists (which is one reason why only company sites adjacent to urban areas have EIPs). In others, it may be very difficult to generate sufficient local interest to form a viable CLC, particularly when that community lacks the educational, technical or cultural resources to communicate

---

29 Few participants felt that consultation involved significant sacrifices, and its benefits were generally seen as far greater than its costs and risks. The major cost involved was the participants' time, especially at the beginning of the consultation process, when the parties were involved in preparation of an initial EIP for the industrial plant concerned. Time requirements appeared to be the major deterrent to greater community participation.

30 This interpretation of the survey results is consistent with the views of the NRC (1997) that an improved understanding of the social context within which others view risks (in this case, the risks stemming from emissions from industrial plants) can help to integrate the different world-views of polluters and pollution sufferers to the extent that each is more willing to accept information provided by, and recognise the legitimate rights of, the other.

effectively with industry.[31] One respondent put this issue into stark relief when he noted the following:

> when I attended the [CLC] meeting, there wasn't any real consultation going on—it was more of a lecture. Most of the community participants had a non-English speaking background, and they were having great difficulty following proceedings. The company representatives sat in a row, and didn't give them any real opportunity to contribute. You could see the frustration building up [in the community members].

It is also possible (though, as far as we could determine, rare) that community members participating in CLCs do not reflect broader community interests but rather the personal agenda of the particular individuals involved. Should this issue arise, it may not be easily resolved given the potential scarcity of community members willing to participate.

Even where a successful CLC is formed in the early stages of an EIP, there is no guarantee that it will continue to function effectively in the longer term. Perhaps the greatest challenge is maintaining community interest in the face of EIP success, for, as the very issues that sparked community involvement are progressively resolved, community engagement is likely to wane. By their very success, CLCs sow the seeds of their own decline. As one respondent pointed out: 'The community consultation committee is a positive way to relate to local people. They had a strong interest to start with but now it's difficult to keep them going. The stirrers have had their say and lost interest.' Another reported that: 'At one of the first EIP meetings 90–100 residents turned up. Now there are about 5–6 regulars. There has been less community pressure because there have been real improvements.' Even worse, in some cases interest can drop to a level where the entire process ceases to be viable and the meetings must be abandoned for want of a community presence. Community engagement may also wane if the participating community members lose confidence in the legitimacy and/or effectiveness of the EIP process.

Yet none of these phenomena is inevitable, and we identified a number of notable exceptions where community groups have continued to make a long-term contribution to the EIP process. Such has been the case at the Altona chemical complex site, where a high level of community concern and participation has been maintained for many years, and at a number of large company sites.

Even where an effective CLC exists, the environmental concerns of local community members may diverge substantially from wider environmental priorities. For example, communities are likely to focus on those activities that are both visible and local in impact, such as noise pollution, odour and particulate air emissions. Yet noise pollution does not have any lasting adverse effect on the environment and odour, while understandably offensive to the local community, is similarly unlikely to have any lasting negative impacts on community or environmental health. Even particulate air emissions, such as dust, commonly have limited environmental impact beyond the immediate vicinity of

---

31   Experience suggests that those that have a higher proportion of educated, middle-class professionals are far more likely to create successful EIP Community Liaison Committees than others. Members of such communities are more likely to have the necessary literacy skills and a greater command of technical concepts that enable them to engage with company representatives on a more equal footing. They are also far more likely to be interested in environmental issues, and willing to act on this interest.

the enterprise premises. The danger inherent in the EIP process is that, by focusing attention on such local community environmental concerns, scarce enterprise resources are diverted away from other less obvious but more significant environmental issues. The latter might include greenhouse gas emissions and energy efficiency, emissions of ozone-depleting substances, polluting effluent entering the sewerage system, the handling of prescribed wastes, or emissions of volatile organic compounds. And, in extreme cases, the 'not in my back yard' (NIMBY) syndrome may produce environmentally counter-productive results.[32] On the other hand, while EIPs commonly started out by focusing almost exclusively on local issues, over time those that did not simply wither away (a minority)[33] tended to become more sophisticated, and to address a much broader range of environmental issues. For example, the Altona CLC has shifted from a narrow focus on basic regulatory compliance and noise and odour pollution to encompass a much more ambitious set of environmental goals.

Those EIPs that do 'last the distance' also tend to evolve in terms of their broader expectations of corporate environmental management practices. That is, over time, the community (often encouraged by the VEPA client manager) may push for more sophisticated monitoring and reporting mechanisms and, significantly, the implementation of process-oriented standards such as EMSs. More commonly, community involvement pushes the enterprise to place a higher priority on environmental issues, and, having done so, the enterprise itself initiates further measures as it shifts from a reactive to a proactive approach. As one industry respondent noted: 'We try to anticipate community concerns, not just wait for objections. It's about building up a relationship . . . and, importantly, giving them [the community members] some control.'

Facilitating industry self-management and environmental partnerships

One of the goals of EIPs and accredited licensing was to ease the regulatory load on VEPA (enabling it to redeploy its scarce resources where they are most needed). This was to be achieved both by empowering communities to act as surrogate regulators and by facilitating greater industry self-management, as companies learned, though developing their action plans and environmental improvement processes, how to better manage their environmental affairs. The shift to industry self-management was also intended to be accompanied by a shift in the relationship between regulated companies and VEPA from adversarialism to partnership. Such a shift was likely to occur in its fullest form in the case of accredited licences, where the obligation to implement an EMS, and to undertake auditing of it, would move the company substantially down the road to self-management.

---

32 For example, there is invariably strong local community opposition to the use of on-site high-temperature incineration for the disposal of hazardous wastes. This is despite the availability of often compelling scientific evidence that this is a highly effective overall environmental solution.

33 Many of the earlier EIP participants are into their fourth and fifth version, and these have tended to become more sophisticated over time. At any one time, a number of EIPs will be up for renewal. At the moment, seven existing EIPs are in preparation for a new version. If we include those EIPs that are no older than three years, this means that 32 are either in the process of being renegotiated or relatively young. This leaves around 20 that are four years old or older.

However, even the process-based requirements of the EIP were a substantial shift in this direction.

In practice, there has so far been little or no saving of VEPA resources as a result of the EIP process. For those enterprises that had a poor environmental record prior to the introduction of EIPs, there has been no substantial change in regulatory direction. VEPA officials broadly agreed that community pressure and oversight were making a positive contribution, but took the view that this alone was insufficient and that only by 'keeping the pressure on' would environmental improvement be assured. For this reason they maintained a regulatory underpinning by including, or threatening to include, EIP environmental improvement targets as part of a company's licence conditions, and threatening and invoking sanctions in the event of breach. Thus there remains a heavy emphasis on traditional inspection and enforcement to achieve compliance.

For those who adopted an EIP voluntarily (i.e. good performers) and even more so for accredited licensees, VEPA inspectors were far more likely to employ a 'regulatory light touch'. However, it was far from obvious that this situation was substantially different from the one that might have existed previously. In other words, VEPA officials in any event, and for understandable reasons, subjected environmental leaders to less regulatory scrutiny than laggards, and the degree of scrutiny does not appear to have changed markedly in the case of those adopting EIPs (or accredited licences). Thus, it does not appear that the EIP or accredited licensing processes have brought about in any substantial redeployment or saving of regulatory resources.

Regulated enterprises, on the other hand, may have achieved resource savings by virtue of the greater flexibility provided to them under the EIP process, and even more so under accredited licensing. However, few enterprises in our sample placed emphasis on this aspect and none indicated that they had enrolled in either of these initiatives for this reason. This may in part reflect the emphasis of EIPs, at least in their early phases, on environmental issues with a local impact, such as noise, odour and particulates. Addressing these issues is unlikely to result in any resource savings at all. The only resource issue raised by respondents was in relation to some of their efforts to address energy efficiency in later iterations of their EIPs. Nor did they suggest that these mechanisms gave them greater opportunities to find win–win solutions: for example, through cleaner-production initiatives. Indeed, several respondents suggested that textbook cleaner-production win–win outcomes are not widespread, and that, on the contrary, most environmental improvements under EIPs and accredited licensing have been very costly to achieve. For example, one respondent stated: 'Meeting our EIP targets hasn't been cheap. Even though the reduction in licence fees [from an accredited licence] has been welcome, there is no way that this has offset the costs of our improvements.'

A related aspiration of VEPA in introducing EIPs and accredited licensing was to provide an alternative to command-and-control regulation which would result in the formation of 'strategic alliances and forming or facilitating partnerships' (VEPA 1993a). The partnerships to which they refer are principally those between the VEPA and participating enterprises, and were intended to bring about a more co-operative and less adversarial relationship between the two parties. Since the large majority of our industry respondents were very supportive of the EIP and (to some extent) the accredited licensing initiatives, one might have anticipated an equally positive change in industry–VEPA relationships. Yet, on the contrary, a high proportion of respondents, whether VEPA or company representatives, were at best sanguine and at worst dismissive about the possibilities of achieving a constructive change in VEPA–company relations in general and about the prospect of 'environmental partnerships' in particular.

A considerable number of industry respondents considered that VEPA, and its client managers in particular, had contributed little to the EIP and accredited licensing implementation processes. First, many felt that VEPA had not provided much support to enterprises, particularly during the drafting and development phase. Enterprise representatives felt that they had been 'effectively left to their own devices' in determining the shape and content of their EIP. When it came to the establishment and maintenance of CLCs, they also considered that VEPA support had been less than forthcoming. Second, the nature of their relationship with VEPA was perceived to have changed very little, with no greater degree of trust or co-operation having developed in regulatory practice. As one enterprise respondent described it:

> Although the EIP has improved some aspects of our relationship [with the VEPA], I would say that overall there has been little change. We are still seen as big and ugly . . . There is no such thing as a lighter regulatory touch—they still treat us the same as before.

This view was not restricted to enterprises that had been compelled, either by community conflict or by VEPA pressure, to adopt EIPs, but also extended to companies arguably categorised as environmental leaders. For example, one accredited licensee claimed that:

> The EPA is terribly inflexible . . . it depends a lot on the individual. I haven't noticed a change in our relationship [after the receipt of an accredited licence]. A different EPA officer came out and criticised our communication processes. [They were] just trying to be a policeman.

The reason for the lack of progress towards a partnership approach, despite goodwill on both sides towards the EIP and accredited licensing processes, can in large part be explained through appreciating the nature of the relationship between industry, VEPA and the community. This was encapsulated by one respondent as follows: 'The EPA is caught between a rock and a hard place. They may want to be more co-operative, but the community won't let them. The community doesn't recognise the benefits of accredited licences.'

Thus the available evidence suggests that, while EIPs, in particular, have been substantially successful in terms of achieving improved environmental performance on the part of industry and in terms of community dialogue and empowerment (albeit only in a limited range of circumstances), they have been far less successful as a form of facilitative regulation, or in generating a partnership approach in industry–VEPA relations. Nor have such partnerships evolved in terms of accredited licensing.

Discussion and policy implications

What are the broader lessons that can be learned from the empirical experience with EIPs and accredited licensing? To what extent, and in what ways, can these initiatives be built on and lessons learned that are applicable to regulatory flexibility initiatives internationally? As we have seen, EIPs have been considerably more successful than accredited licences, and the two mechanisms must be addressed separately.

EIPs: integrating process, outcomes and oversight

EIPs have been largely successful in achieving improved environmental outcomes on the part of participating enterprises. Certainly, the changes are more deep-seated in the case of volunteers, and it is only this group that has to any extent been persuaded to go beyond compliance with existing regulatory requirements. Nevertheless, EIPs have also been successful in making conscripts more conscious of their environmental obligations and how to discharge them, and in conjunction with community and regulatory pressure, in leveraging improved environmental outcomes.

The key to the success of EIPs lies in interactions between, and the mutually reinforcing nature of, the EIPs' central components. EIPs emphasise a process-based approach to environmental protection, through which industry is encouraged to examine systematically its environmental impact and means of reducing it. While proponents believe that this approach will succeed in generating self-organisation and in internalising environmental change, critics fear that it will tempt regulatees simply to go through the motions, to pay lip service to the process in order to meet their EIP requirements and 'keep the EPA off their backs'.[34] The danger will be all the greater because EIPs, exceptionally, are targeted not just at best performers that voluntarily aspire to beyond-compliance outcomes but also at environmental laggards that would not choose to adopt EIP requirements voluntarily. Arguably, it is heroic to impose such requirements on conscripts at a time when the jury is still out as to their efficacy, even in the case of volunteers.[35] The fact that such tokenism was largely avoided can be attributed to the effectiveness of the other two pillars of the EIP strategy: performance standards and community and VEPA oversight.

As we have seen, a number of environmental targets are negotiated with the community and VEPA as part of the CLC process. These targets are clearly identified and transparent to all parties. Moreover, the company itself must engage in regular self-monitoring and report progress in terms of achieving these targets. Accordingly, there is very little likelihood that backsliding and regression into tokenism will go unnoticed. In turn, the identification of targets facilitates the final component of EIPs' success: community and VEPA oversight and enforcement. It is largely because there are benchmarks against which to measure success or failure that both are able to exert effective and continuing pressure. These, too, work in tandem: if community pressure and shaming are insufficient to bring a company back on track, VEPA can seek to modify the licence specifications and threaten enforcement and sanctions. Usually the mere threat to do so will be sufficient incentive to action. While this latter threat is barely relevant in the case of volunteers, for conscripts it may be crucial.

However, while EIPs have been substantially successful in achieving their policy goals, we have also identified a number of respects in which their success is qualified and where

34 For an excellent overview of the issues, see Coglianese and Nash 2001.
35 See e.g. 'Environmental Management Systems: Paper Tiger or Powerful Tool?', www.ivf.se/industrial_environment/; and www.eli.org/pilots.htm and www.environmental-performance.org.

there is scope for further improvement. First, there are sometimes problems in communicating the community and VEPA message to senior management. The result may be that there is a substantial gap between the level of commitment at the plant (where community pressure and EIP have most direct impact) and senior decision-makers at head office. This is a serious concern, for without senior management commitment it will be extremely difficult to embed environmental considerations more deeply within the corporate culture and to achieve deep compliance.

There are a number of strategies that might be successful in increasing the overall impact of the EIP by gaining greater commitment from senior management. In particular, the gap between plant and central management might be substantially reduced if commitments made in EIP negotiations and agreements are made binding on the company as a whole and on senior corporate management in particular (which is not necessarily the case at present).[36] The most obvious way to do this is to ensure that such agreements are signed at board and CEO level (something that is already required in Western Australia).[37] Of course, corporate management will be reluctant to sign legally binding undertakings but, in the context of quasi-compulsory EIPs, the enterprise has little choice in the matter, given EPAs' extensive statutory powers and capacity for arm-twisting. And, in the case of voluntary EIPs, the company will presumably already have determined to set out on a path of better community relations (through better environmental performance), which it will find difficult to achieve in the absence of firm and binding commitments.

Another, related means of increasing senior management commitment would be to place greater emphasis on personal liability, since most surveys confirm that this is the single most powerful motivator of senior management commitment (Henriques and Sadorsky 1995; McCloskey and Maddock 1994; KPMG 1996). A greater emphasis on due diligence and substantial penalties for individuals in the event of serious breach (which would flow directly from signing off the EIP at CEO level) would do much to underpin and reinforce the more constructive process-based emphasis of EIPs, not only in the case of incompetents and the boundedly rational, but also in the case of recalcitrants. As one environmental manager confirmed, 'due diligence and legal liability of directors and officers . . . has certainly made directors more aware of the issues, and more willing to move to widespread use of EMS'.

Finally, senior management commitment might be reinforced if senior managers were required under the terms of the EIP to attend at least some CLC meetings. This already occurs in some enterprises and, according to our respondents, is very effective in focusing those managers' attention directly on the issues. Perhaps more important, environmental managers reported that it commonly gave them considerably greater leverage on environmental matters. This is particularly important since it is often the environmental manager who plays a key role in managing environmental change. He or she is generally 'the crucial medium who translates messages from the external environment (for example from the community and government), communicates them effectively within the firm, and facilitates the internal dialogue which necessarily precedes change' (Parker 1999a: 224).

36 At present, EIPs can be signed off at a variety of levels, which may, but need not, include the CEO, and might only involve the plant manager.

37 Here, in much the same way as in New South Wales, it is the CEO who, under the Environmental Offences and Penalties Act, 1997, must sign off on various corporate undertakings in the licence.

A second limitation of the current EIP process relates to community participation. CLCs have been very successful, not only in increasing pressure on enterprises to improve their environmental performance but also in greatly improving relationships between communities and companies. However, this success is qualified insofar as CLCs operate effectively only where a local community exists and has the skills and motivation to engage in dialogue and oversight. For these reasons, EIPs are to be found almost exclusively in relation to large, highly visible companies operating in close proximity to urban populations. Even here, some committees experience 'burn-out', or are victims of their own success and ultimately decline or disband.

How might the considerable benefits of EIPs be extended to enterprises willing and able to introduce them, and yet unable to garner a sufficient level of community interest or continuing involvement? Where the problem is not the lack of a relevant community, but its unwillingness to join a CLC, one option is to pay community EIP participants to participate, which is precisely what one company in our sample has done. However, although this has been effective in enticing community participation, it does raise broader questions about the credibility of its EIP and the independence and impartiality of community members. At present, there is insufficient evidence to know whether the risks of capture and co-option in these circumstances outweigh the benefits of community participation, or whether this practice should be encouraged, or even condoned, in the future.

A more ambitious alternative to local community participation would be to invite representatives of a range of state and/or national environmental organisations to fulfil the roles of community dialogue and oversight. By doing so, it might be possible to extend the application of EIPs to enterprise premises located in isolated areas and/or to enterprises whose size, or level of environmental impact, are insufficient to attract community attention. Representatives of environmental organisations might also bring to the table a higher level of expertise and sophistication than that possessed by the average CLC member. However, in the past environmental organisations have kept their distance from industry and VEPA in order to maintain a more credible and critical independent voice.[38] From the corporate side, there may be equal discomfort with such an arrangement. Indeed, many corporate managers prefer not to deal directly with environmental organisations, and some consider their objectives to be incompatible. For these reasons, enterprises may be very reluctant to disclose details about their environmental management practices and performance for fear that this information will be used against them at some future time by the environmental organisation concerned.

Notwithstanding the considerable obstacles to co-operation between environmental NGOs and industry, in recent years there has been some considerable thawing in relations between them. For example, an increasing number of enterprises and environmental NGOs have entered into environmental partnerships, having found through bitter experience that co-operation is ultimately a more productive means of securing at least some objectives than conflict (see e.g. Murphy and Bendell 1997). The parallel trend towards corporate environmental reporting also suggests a greater willingness on the part of

---

38   This has certainly been the stance that major environmental groups have taken with regard to participation in the chemical industry's Responsible Care initiative, and in consequence none has agreed to participate in Responsible Care's National Community Advisory Panel (which in some ways fulfils a similar role to the one we propose above).

business to become more transparent and to enter into dialogue with a broad range of stakeholders, including NGOs. As a result of these developments, the time may be ripe for experimentation. This might involve the participation of NGOs directly in the EIP process, or, borrowing from the chemical industry's Responsible Care initiative, it might be possible to set up a panel comparable to the National Community Advisory Panel (NCAP), drawn from a cross-section of individual community leaders with particular concerns for environmental safety and health issues.[39] Under either model, some mechanism would be needed to take account of the drain on resources that such a commitment would entail.[40]

Another means of expanding the scope for EIPs would be to facilitate their application to a number of geographically unrelated sites, rather than restricting them to a single facility. Thus a single enterprise operating a number of sites could be allowed to develop a single EIP encompassing all those sites. This would have the advantage of providing economies of scale, engaging management at a more senior level, fostering a more sophisticated management of total environmental impact, and, potentially, engaging some industrial sites not normally brought under the EIP umbrella. In terms of maintaining a representative local community input, community members could be drawn from the range of contributing site localities, and CLC meetings could be rotated through the same site locations. The practice of incorporating multiple sites under a single EIP umbrella has already begun to occur, albeit in an ad hoc fashion.[41] However, a degree of ambivalence was expressed by several industry respondents about the prospect of multiple sites. Although they acknowledged potential gains through economies of scale and uniformity, there was concern that this could detract from on-site commitment and undermine the role of plant managers.

Finally, the reach of EIPs might be extended to address environmental problems at a sub-regional level. This prospect is currently being explored in Victoria, through the vehicle of 'neighbourhood EIPs'. The intention is to provide a 'statutory tool to allow those contributing to and those affected by local environmental problems to come together in a constructive forum'. It is anticipated that participants in such fora will include residents, industry (including SMEs) and local government. As under conventional EIPs, the tools to achieve change would be the CLC, process-based change and environmental target-setting. But, unlike conventional EIPs, the focus would be not a single site but a number of sites (including, for the first time, SMEs) within a particular geographic area and, presumably, non-industrial sources of pollution as well. Interest-

---

39 The NCAP is intended to provide a vehicle through which the public may play an integral role in shaping the Responsible Care initiative. NCAP members review proposed codes of practice from a public-interest perspective, and provide the relevant industry association (PACIA) with feedback on other emerging issues of public concern (Gunningham 1995).

40 It may be unreasonable to expect an environmental organisation, acting in a purely voluntary capacity, to fund the entire cost of its participation. As such, some means of external supplementation may be warranted. As noted above, at least one EIP company is already paying community members to participate in its Community Liaison Committee. While a similar practice could be adopted to facilitate NGO participation, it raises the same concerns about capture and co-option. Such concerns might at least be mitigated if VEPA paid the direct costs of such participation (perhaps by a levy on participating EIP companies); otherwise this might place further pressure on an already very limited regulatory budget.

41 For example, Qenos has four sites included under its single EIP, and Melbourne Water has a single EIP covering its entire city-wide operations.

ingly, the legislation provides VEPA with discretionary powers to mandate the formation of neighbourhood EIPs, raising the question of how they might effectively enforce participation. At the time of writing, this development is only in its embryonic stage, and cannot be further addressed here.

Accredited licensing

Accredited and best-practice licences have so far attracted very few participants. The majority of these have been motivated by a desire to gain credit for internal changes that they have already made for entirely other reasons. For this reason, in their present form accredited and best-practice licences make only a marginal difference to corporate environmental performance, although they do provide greater flexibility and some cost and resource savings to industry, and reward some best performers for their past achievements.

This outcome falls far short of accredited licensing's original aspirations, central to which was the expectation that participating enterprises would 'demonstrate a high level of environmental performance and *an ongoing capacity to maintain and improve this performance*' (emphasis added) (VEPA 1998: 1). Rather than simply rewarding already very good environmental performers for their past achievements, they should be seeking out and attracting new recruits that have not yet achieved best-practice environmental performance, but with encouragement might do so. Many such enterprises would probably come from the ranks of those that had already voluntarily adopted an EIP (i.e. already good performers), and could be encouraged to take the next step of developing an EMS. In practice, it would involve seeking out not merely the 'cream of the cream' (perhaps the top 5%) and rewarding them for their existing achievements, but also attracting the 10%–15% below that, and facilitating their achievement of demonstrable improvements beyond 'business as usual'.

To achieve such a progression, two things are needed: greater incentives to join, and a relaxation in the eligibility requirements so that those that are not now, but might potentially become, top environmental performers are encouraged to participate. Unfortunately, both of these conditions are highly problematic.

The issue of incentives may well be an intractable problem. Accredited licensing may involve substantial transaction costs for enterprises. For example, investment of staff time can be considerable, the negotiation period may be protracted, and facilities may receive conflicting signals from different levels of regulatory staff (as reported by some company representatives). The costs of participation, in other words, may outweigh the incentives. This indeed seems to be the case, with the result that only a minority (most of whom have established the main cornerstone of accredited licensing already) see sufficient value in enrolling. This has also been a recurrent problem in American regulatory flexibility initiatives.[42] But creating additional incentives is no easy task, particularly if the

---

42 For example, the transaction costs in formulating Intel's Project XL programme (allowing Intel to change processes and emissions without obtaining a new permit as long as the company keeps its emissions below levels set by federal regulations) were higher than that of a traditional air permit for all the participants—government agencies, Intel and public volunteers. However, research also found that overall costs are likely to be lower than in the case of traditional regulation since a traditional permit requires the manufacturer to notify the government each time a production change occurs (Boyd *et al.* 1998).

new target group includes those that have not yet gone through the expensive and time-consuming process of developing an environmental management system and of subsequently having it properly audited. The current range of incentives provided by VEPA is already substantial,[43] and the scope to create substantial new ones is limited.

The most plausible options are as follows. First, VEPA could seek to provide much more substantial reputation advantages to accredited licensees. This might involve establishing and promoting a logo and otherwise clearly identifying publicly those who have enrolled and what this signifies, though whether this in itself would entice many new participants is doubtful. Certainly the American experience has been that 'most companies focus on bottom line gains and seem unimpressed by soft incentives such as recognition and assistance—especially compared to the resources required to apply and participate in a program' (Crow 2000: 24). Second, government and government contractors might be required to purchase from (or at least give a purchasing preference to, all other things being equal) accredited enterprises by virtue of 'affirmative procurement' policies such as have been introduced on a modest scale in the United States. However, there would be substantial political opposition to such a suggestion, which also arguably conflicts with broader competition principles.

Third, the role of negative incentives is equally, if not more, important, but these have been relatively neglected in most Australian jurisdictions. For example, there is considerable evidence internationally that the single most important factor in persuading firms to adopt a process-based approach is fear of a less desirable alternative: specifically, tougher enforcement and penalties under traditional regulation. Yet, within Australia, most jurisdictions (because of very limited resources and, in some cases, the prevailing philosophy) either lack a strong commitment to enforcement against wilful polluters or are hampered by an inadequate judicial response when they do prosecute (Briody and Prenzler 1998). Again, these are problems for which there is no ready solution.

Relaxing the eligibility requirements for participating in accredited licensing would also be challenging.[44] Some community and environmental NGOs already view such 'regulatory flexibility' initiatives with suspicion, fearing regulatory capture and a 'deal' with industry that enables it to avoid some of its regulatory responsibilities, with the result that public-interest objectives will be compromised. As we saw earlier, such concerns have seriously constrained the development of a partnership approach between industry and VEPA. If VEPA relaxes its standards to accommodate participants with less than exemplary environmental records, this may inflame community concerns. If one or more of those firms then abuses its privileges in a manner that is visible to public-interest groups, or there is a major environmental incident involving such a company, then VEPA's credibility and the credibility of the programme as a whole is on the line. Small wonder

---

43   Cf. e.g. the list provided in Gunningham and Johnstone 1999.

44   However, as Crow (2000: 24) has noted, in the USA: 'less than half of [environmental leadership programmes] require any proven pollution decrease (or other environmental outcome) either for entry or for staying in the program. Most only require commitments or goals for continual improvement—apparently without penalty for failing to achieve them. This lack of performance requirement may be explained by a key premise common to many programs: that sound EMSs will really ensure continued compliance and, eventually, beyond compliance performance.'

that in these circumstances VEPA is itself split as to the virtue of expanding accredited licensing to a wider group of participants.[45]

There are other opportunities for enhancing the contribution of accredited and best-practice licensing, but none of these goes to its core functions. First, even though the overall number of potential accredited licences might be limited by the number of appropriately large industrial sites (that is, scheduled premises) willing to participate, their reach could be extended down the supply chain by building in the additional concept of product stewardship—and an obligation for participating enterprises to take greater responsibility for the environmental credentials and behaviour of those (much smaller) enterprises with whom they deal or who use their products. That is, large accredited licensees could be encouraged to take on an environmental 'chaperone' role in respect of their smaller commercial customers by facilitating, informing and cajoling them into adopting better environmental management practices. This could include, for example, training programmes and the use of simplified EMSs. Whether the disadvantages of this proposal (providing a further disincentive for participation) would outweigh the advantages, only experience will tell.

A more ambitious initiative would be to integrate accredited licences into the top tier of a broader information regulation strategy along the lines described in Chapter 6: for example, using the PROPER programme as a model. This would have the considerable attraction of facilitating NGOs and the community more generally in identifying worst performers and free-riders, while also giving best performers greater public relations rewards. This would, however, involve an ambitious new initiative of substantial proportions, and a major investment in 'informational regulation'. It would also require the capacity to make meaningful comparisons between firms, in order to know which ones to rank in each category. Some form of environmental accounting and measurement would be essential, in a form that allows for a qualitative measure of the environmental performance of the participants in the programme. This is a considerable challenge, given that reliable and transparent data sources are not developed in Australian jurisdictions.[46]

---

45 In the majority were those, such as the following VEPA official, who expressed the view that accredited licences should be reserved for 'only the best environmental performers' and 'those that have an exemplary track record'. However, there were others, such as the following official, who felt that: 'they [accredited licences] are there to provide an inspiration for all scheduled premises—everyone can see the benefits, and everyone can try to make the changes necessary for approval'.

46 The Green Zia Program of New Mexico, USA, has been identified as a possible model for such performance-based measures. According to a Harvard University Report: 'the Green Zia Program measures environmental performance of the facilities that use the metric based on a proven quality model (i.e. Malcolm Baldrige National Quality Performance Award Program). Trained, independent examiners are used to score the written program applications using a rigorous scoring method. Feedback reports are provided to the applicants to use in their mandatory continuous improvement effort. The scores from individual facilities can be aggregated by industrial sector or geographical area to show how the . . . Program is improving environmental performance without having to list a large number of uncoordinated environmental indicator metrics. The environmental results featured in the applications can be used to assess the benefits of the Program and report them to the public on an aggregate basis' (Harvard School of Public Health 2000).

Given the very substantial challenges in improving the effectiveness of accredited licensing, and its disappointing record in its early stages, it is questionable whether it justifies the costs involved. There are costs in developing such a licensing scheme and in negotiating each individual flexible licence (even if the onus is placed on proponents to demonstrate that a proposal would meet or exceed the regulatory objectives). If these costs are substantial, they may result in the redeployment of regulatory resources away from laggards and towards the sort of leaders who propose regulatory flexibility initiatives, resulting in a serious resource misallocation.[47] These costs might well be offset if VEPA were able to rely largely on 'industry self-management' by accredited licensees, redeploying resources previously used in regulating such licensees to other (and better) purposes. Unfortunately, the limited evidence available suggests that very little redeployment of regulatory resources has occurred.[48] Our preliminary research suggests that any cost savings so far are insubstantial.

Conclusion

The story of Victoria's two closely related, but conceptually distinct, facilitative regulation initiatives contains a number of lessons for the future of regulatory reform. One the one hand, EIPs have been a very considerable success. Not least, they have directly empowered local communities, increased pressure on companies to improve their environmental performance and, through structured dialogue, greatly improved relationships between communities, enterprises and VEPA. In so doing, they have improved environmental outcomes on issues of community concern, increased the level of trust between communities and companies and created a more predictable investment environment.

The key to this success lies in the interconnections between, and mutually reinforcing nature of, their central components. Process-based requirements which prompted companies to develop more effective internal management and planning, community participation and oversight, agreed environmental targets and the threat of direct VEPA intervention, all contributed to achieving substantially improved environmental outcomes, and the absence of any one of these components would have seriously weakened the overall strategy. While the success of this approach in dealing with environmental laggards was more qualified than in dealing with leaders (the changes were less deeply embedded and only effective in conjunction with VEPA pressure), even here the evidence suggests a substantial improvement on the status quo.

But perhaps their greatest contribution has been in institutionalising community dialogue and participation and in empowering communities to act as environmental 'watchdogs', and as surrogate regulators, at least in circumstances where they are ready, willing and able to take on these roles. While community participation has been confined to large enterprises located in urban areas with a concentration of local residents, there

---

47 This point is made in broader terms in Commission for Environmental Co-operation 1998.
48 We described earlier how regulators were reluctant to change their practices in the case of environmental leaders that adopted an EIP, and there is no evidence of any greater resource shift in relation to accredited licensees.

is scope to expand their use to a broader range of circumstances. This might be achieved: by recruiting environmental NGOs to play a similar role where a community is unwilling or unable to do so; by expanding a single EIP to a number of corporate sites; by developing neighbourhood EIPs; or through a combination of these approaches.

However, not all of VEPA's goals in introducing EIPs have been achieved. There is very little evidence to suggest that EIPs have provided significant regulatory resource savings. Nor have EIPs achieved the sort of transition from adversarialism to partnership in relations between VEPA and EIP participating companies. The reasons for this may be largely structural, suggesting that there is only very limited scope for change in the face of communities that are very distrustful, or for more harmonious relationships between regulators and regulatees. But what they have achieved, both in terms of improved environmental performance and improved company–community relations, is considerable.

In contrast, accredited licensing has delivered very little of what it promised. In its present form, its main virtue is to provide greater flexibility and some public relations benefits to participants, in circumstances where they have already established a systems-based approach to environmental protection, but there is little evidence of overall environmental improvement resulting from this initiative or its Western Australian counterpart. Considerable redesign would be necessary if this form of regulatory flexibility is to fulfil its broader policy aspirations. Unfortunately, such redesign (providing more substantial incentives, and broadening the range of participants) will not be easy. Indeed, given the severity of the obstacles it confronts, the uncertainty of the outcome, and the paucity of cost savings available to regulators, it is doubtful that it is worthwhile. If so, then accredited licensing, like many of its United States counterparts (Crow 2000), seems destined to represent not a paradigm shift but a very modest tinkering around the edges of the current regulatory regime.

CONCLUSION
Reconfiguring environmental regulation

This book has explored the changing role of the regulatory state, and the evolution of a number of next-generation policy instruments intended to overcome, or at least to mitigate, the considerable problems associated with previous policy initiatives, and traditional forms of regulation in particular. The goal is, in the words of the United States EPA (2000a: 1), 'to adapt, improve and expand the diversity of our environmental strategies' and to address the circumstances not only of laggards but also of leaders.

However, policy reform has taken place in what is, in many respects, a hostile political and economic environment. The 1980s and 1990s saw a resurgence of free-market ideology which, assisted by the economic and political collapse of the former Soviet Union, enabled neoliberalism to triumph almost unchallenged, for most of that period and beyond. And, while public opposition precluded the sort of wholesale deregulation that occurred in some other areas of social regulation, environmental regulatory budgets were substantially cut in almost all jurisdictions. This trend shows little sign of changing under the lower-taxation regimes that now characterise the large majority of economically advanced states, irrespective of the party in power.

During the same period, governments have also experienced considerable pressure from industry to reduce the economic burden of complying with environmental regulation. Although, on most calculations, the costs of compliance are relatively modest,[1] industrial lobby groups have argued strongly, and often successfully, that the imposition of such regulation would put industry at a competitive disadvantage. In an era of globalisation, in which capital flight to low-tax, low-regulation regimes is increasingly plausible (though far less often demonstrated), governments have listened particularly closely to industry concerns and have frequently responded sympathetically. Thus the confluence of economic and political pressures has often precluded the application of direct regulation.

But, while government regulators have been losing both their power and resources, others have begun to fill the regulatory space they previously occupied. For example, envi-

---

1   On most calculations, these are between 1% and 2% of GDP.

ronmental NGOs, aided by advanced techniques for information-gathering (from digital cameras to satellite imaging) have become increasingly sophisticated at communicating their message (via global television, international newspapers and the Internet) and in using the media (and sometimes the courts) to amplify the impact of their direct-action campaigns. They have not only sought to shape public opinion to lobby governments and to pressure industry directly but also to influence consumers and markets through strategies such as orchestrating consumer boycotts or preferences for green products. Indeed, they have commonly bypassed governments altogether where they perceived them to be overly sympathetic to industry or incapable of effective action.

At the same time, a variety of commercial third parties have also begun to take a far greater interest in environmental issues. Banks and insurance companies seek to minimise their financial risk by scrutinising more closely the environmental credentials of their clients (Schmidheiny and Zorraquín 1996). And financial markets themselves have become responsive to good or bad environmental news, rewarding environmental leaders with a share price increase and discounting the share price of laggards (Lanoie *et al.* 1997). So too is supply chain pressure increasingly important, with a substantial number of companies seeking accreditation under ISO 14001—not because regulators require it or because they believe it necessary but rather because their trading partners insist on it.

As part of this reshaping of the regulatory landscape, a number of environmental stakeholders have to some extent departed from their traditional roles. Some business groups, such as the World Business Council for Sustainable Development, have become proactive, arguing that business is part of the solution rather than merely the problem, and sought to develop a variety of voluntary initiatives through which business seeks to shape its own environmental destiny. Environmental NGOs, frustrated with their limited impact on governments, or at the ineffectiveness of government in protecting the environment, have redirected their attention towards corporations through strategies ranging from confrontation to partnership. And government policy-makers, constrained by diminishing resources, and noting the increasing power of NGOs and financial markets, and the potential for industry self-management, have became increasingly enamoured with the possibilities of 'steering not rowing' in policy design.

As we have seen, what has evolved is not a retreat of the regulatory state and a return to free markets but rather a regulatory reconfiguration. The United States EPA's Reinventing Environmental Regulation programme, negotiated agreements in Western Europe, a plethora of informational regulation initiatives, various forms of industry self-management and a variety of enterprises (commonly using supply chain and financial market pressure) built around harnessing third parties as surrogate regulators, nevertheless involve a continuing government role. Even in relation to problems that the state is ill equipped to address directly, it almost invariably retains a supporting role, underpinning alternative solutions and providing a backdrop without which other, more flexible, options would lack credibility, and stepping in where they fail. That is, in almost all circumstances the state is still involved in engineering solutions to environmental problems rather than trusting the market, unaided, to provide them.

This reconfiguration is still in process, and the next-generation instruments that have emerged are very diverse. Some seek out and nurture win–win solutions, some seek to replace conflict with co-operation between major stakeholders, and others seek to mitigate power imbalances, and to increase transparency and accountability, as is the case with informational regulation. Many, in stark contrast to the first generation of command-and-control, seek to encourage and reward enterprises for going beyond compliance with existing regulation—to reward leaders rather than merely to drag laggards up to a minimum legal standard. But neither the precise direction of this reconfiguration nor its results are yet known. Much work remains to be done in mapping progress, identifying what works and what doesn't, and why, and in providing a better understanding of how to match types of instrument, and institution, with particular environmental problems. Our earlier analysis and case studies are intended to contribute to this enterprise.

In the following sections, we provide a broader perspective on this regulatory reconfiguration. First, we examine it through a variety of different lenses and in terms of a number of different conceptual frameworks. Second, we reflect on some broader lessons for the future of regulatory reform.

Conceptualising regulatory reconfiguration

Below we examine five different frameworks, or lenses, through which one might better understand regulatory reconfiguration. None of these lenses offers (or necessarily purports to offer) a complete prescription for what the next generation of policy instruments should involve. However, as we will show, individually and collectively they enrich our understanding of the individual policy instruments we have examined in previous chapters, and what they might achieve. They also provide insights into the challenges facing regulatory reconfiguration and how they might be resolved.

Reflexive regulation

The literature on reflexive law[2] recognises that the capacity of the regulatory state to deal with increasingly complex social issues has declined dramatically. As Teubner (1983) and others (Teubner et al. 1994) have argued, there is a limit to the extent to which it is possible to add more and more specific prescriptions without this resulting in counterproductive regulatory overload. Traditional command-and-control regulation (a form of 'material law')[3] is seen as unresponsive to the demands of the enterprise and unable to generate sufficient knowledge to function efficiently. In sum: 'the complexity of society outgrows the possibilities of the legal system to shape the complexity into a form fitting to the goal-seeking direct use of law' (Koch and Nielsen 1996). To give a concrete example, one cause of the Three Mile Island (TMI) nuclear accident and near meltdown was that operators simply followed rules, without any capacity for strategic thinking, and, as

2   This, in Teubner's (1983) terminology, is a form of law using indirect means to achieve broad social goals, being a distinct shift from the previous approach (material law), which has broad goals but uses specific direct means to achieve these goals. Material law, according to Teubner, has largely failed because the complexity of modern society is incapable of being matched by a legal system of comparable complexity capable of harnessing direct, goal-seeking law to accommodate social goals.

3   Defined as a form of law having broad goals and using specific direct means to achieve the goals.

events unfolded that were not covered by a rule, they had no capacity to read the situation and respond appropriately.[4]

In contrast, reflexive[5] regulation, which uses *indirect* means to achieve broad social goals, has, according to its proponents, a much greater capacity to come to terms with increasingly complex social arrangements. This is because it:

> focuses on enhancing the self-referential capacities of social systems and institutions outside the legal system, rather than direct intervention of the legal system itself through its agencies, highly detailed statutes, or delegation of great powers to the courts . . . [it] aims to establish self-reflective processes within businesses to encourage creative, critical, and continual thinking about how to minimise . . . harms and maximise . . . benefits (Orts 1995: 1,232).

Put differently, reflexive regulation is procedure-oriented rather than focused directly on a prescribed goal, and seeks to design self-regulating social systems by establishing norms of organisation and procedure (Fiorino 1999).

Such a strategy can also be viewed as a form of 'meta risk management' whereby government, rather than regulating directly, risk-manages the risk management of individual enterprises.[6] This is what happens under the 'safety case' regime, instituted on North Sea oil rigs following the Cullen enquiry into the Piper Alpha disaster where 167 lives were lost (Cullen 1990). This involves what is in effect a safety management system being developed by the rig operator and submitted to the regulator for scrutiny and approval. Similarly, the safety regime established for the nuclear power industry, post TMI, ceased to be primarily about government inspectors checking compliance with rules, and more about encouraging the industry to put in place safety management systems which were then scrutinised by regulators: in this case, by the industry association in the form of the Institute of Nuclear Power Operations (Braithwaite and Drahos 2000; Rees 1994).

A number of the second-generation instruments examined in previous chapters could be readily interpreted as examples of reflexive law, whose goal, rather than regulating prescriptively, is to encourage organisations to establish processes of internal self-regulation to monitor, control and replace economic activities injurious to the environment.[7] Take the use of environmental management systems, which form the principal component of regulatory flexibility initiatives and some forms of negotiated agreement. Such systems seek by law to stimulate modes of self-organisation within the firm in such a way as to encourage internal self-critical reflection about its environmental performance. As we have seen, they establish processes and procedures that encourage self-reflexive learning and thinking about reducing environmental impact rather than seeking to influence behaviour directly by proscribing certain activities. As such, they represent a quintessential example of reflexive law, as indeed do the EIPs we examined in Chapter 8. Similar mechanisms are being devised to suit the circumstances of SMEs. These include not only 'slimmed-down' EMSs but also self-inspection and self-audits, checklists and the sorts of cleaner production initiative as in the smash repairs industry, described in Chapter 3.

4   For an excellent analysis of the alternative and much more reflexive regime that evolved in the aftermath of TMI, see Rees 1994.
5   The term 'reflexive' derives from Teubner 1983.
6   This concept is further developed by Parker (1999b, 1999c) and in an as yet unpublished work by my colleague, John Braithwaite.
7   For a development of this concept, see Orts 1995.

In part, informational regulation can also be viewed in these terms (although it is much else besides). For example, requiring facilities to track and report their emissions (as under the TRI) not only empowers community groups, and enables markets to make more informed judgements, but it also leads to a degree of self-reflection on how things might be done differently. Dow Chemicals is among those enterprises that freely acknowledges that it did not previously measure its wastes and as a result had no idea how much it was discharging. Once the enterprise did so, it realised that there was a business opportunity to make pollution prevention pay, through re-use, recycling, the substitution of different substances and the use of fewer chemicals (Gunningham and Cornwall 1994). Thus a strategy that involved no requirement to do anything other than estimate discharges and disclose them had a variety of broader consequences, including to generate internal organisational change (and corporate shaming), which in turn resulted in substantially improved environmental performance for many companies.

On close inspection, a number of other strategies also contain elements of reflexive regulation. Industry self-management initiatives certainly fall within this category, to the extent that they deliberately build in a variety of mechanisms to generate internal compliance and self-organisation (as exemplified by INPO, described in Box 4 on page 100). Even economic incentives, on one view, have reflexive elements, though whether their designers would have viewed them in these terms is debatable. Nevertheless, Fiorino (1999: 450) argues that marketable permits such as emissions trading and acid rain allowance trading programmes in the United States 'induce reflection by specifying a goal and allowing firms to decide how to achieve it, given their circumstances'. However, he also notes that because they are implemented in the context of technology requirements such permits involve a combination of substantive and reflexive law.

Regulatory pluralism

Traditionally, regulation was thought of as a bipartite process involving government and business, with the former acting in the role of regulator and the latter as regulatee. However, a substantial body of empirical research reveals that there are a plurality of regulatory forms, that numerous actors influence the behaviour of regulated groups in a variety of complex and subtle ways (Rees 1988: 7), and that mechanisms of informal social control often prove more important than formal ones (Gunningham 1991). In the case of the environment, the regulatory pluralism perspective suggests that we should focus our attention on the influence of: international standards organisations; trading partners and the supply chain; commercial institutions and financial markets; peer pressure and self-regulation through industry associations; internal environmental management systems and culture; and civil society in its myriad forms.

These insights have led some policy-makers to investigate how public agencies may harness institutions and resources residing *outside* the public sector to further policy objectives in specific concrete situations. This approach can be seen as part of the broader transition in the role of governments internationally: from 'rowing the boat to steering it' (Osborne and Gaebler 1992) or choosing to 'regulate at a distance' by acting as facilitators of self- and co-regulation (Grabosky 1995) rather than regulating directly. Thus for regulatory pluralists, environmental policy-making involves government harnessing the capacities of markets, civil society and other institutions to accomplish its policy goals more effectively, with greater social acceptance and at less cost to the state

(Gunningham *et al.* 1999). And, since parties and instruments interact with each other and with state regulation in variety of ways, careful regulatory design will be necessary to ensure that pluralistic policy instruments are mutually reinforcing, rather than being duplicative or, worse, conflicting (Gunningham and Grabosky 1998: Ch. 6).

A substantial number of next-generation instruments are pluralistic in conception. Some, such as the regulatory flexibility initiatives established under the Clinton–Gore Reinventing Environmental Regulation initiative, were directly inspired by one version of regulatory pluralism (and by Osborne and Gaebler's [1992] concept of 'steering not rowing' in particular). Seeking to embed environmental values and processes within the corporate culture in such a way that it becomes self-regulating, and relying on oversight from local communities and perhaps third-party auditors, to supplement or even replace direct regulation, is a quintessential pluralist strategy.

Many informational regulation initiatives can also be understood in pluralist terms. Providing communities and financial markets with greater information about corporate environmental performance effectively empowers both of these groups. Communities and environmental NGOs respond by using this information to shame bad corporate performers, while the same information apparently influences share prices, thereby indirectly punishing bad performers and rewarding environmental leaders (Lanoie *et al.* 1997). In particular, the powerful impact of the TRI as a surrogate regulatory tool is well documented (Fung and O'Rourke 2000).

Many of our case studies provided further concrete illustrations of the power either of industry itself or of commercial and non-commercial third parties to function as surrogate regulators. In the case of the Victorian vegetable growers, supply chain and community pressure played important roles. In the case of motor vehicle smash repairs, the insurance industry's role was pivotal, while in both the ozone protection and mining industry studies industry self-management was the critical instrument, *albeit* with varying degrees of success. And permeating many of our other examples and studies was the role of communities and NGOs as effective de facto regulators, perhaps best illustrated by the history of Environmental Improvement Plans and the role of Community Liaison Committees in Victoria.

Environmental partnerships

Environmental partnerships came of age in the 1990s when parts of industry, government and NGOs recognised that conflict and confrontation were not necessarily the best means of achieving either the best economic or environmental results. Governments sought alternatives to direct regulation, and business enterprises, dissatisfied with the cost and inflexibility of command-and-control regulation, and sometimes seeking win–win outcomes, sought more flexible and less confrontational alternatives. NGOs, too, began to see virtue in 'green alliances' with environmentally proactive enterprises. Sometimes these partnerships involve agreements between business and NGOs, or between governments and business, or even between business and business along the supply chain. On other occasions, they may embrace governments, NGOs, business and a range of other third parties, who, as we have seen, held out the promise of acting as surrogate regulators and performing many of the functions that government regulation was no longer ready, willing and able to fulfil.

According to their proponents, environmental partnerships provide an additional policy option that steers a middle course between the two extremes of traditional regulation on the one hand and self-regulation and voluntarism on the other, and, in so doing, take advantage of their respective attributes while compensating for their particular weaknesses (Long and Arnold 1994). Environmental partnerships also provide opportunities to replace adversarialism with co-operation, and, in doing so, may provide benefits for all sides. For example, through 'green alliances' business may obtain the political goodwill and credibility that NGOs bring to the partnership, while, in return, environmental groups gain commitments to improved environmental practices on the part of their business partner. In industry–government partnerships, governments can offer resources, expertise, regulatory relief and external legitimacy in return for improved industry environmental performance. Government can also play a broader role in encouraging, facilitating, rewarding and shaping a variety of partnership forms.

A number of examples of such partnerships were given in preceding chapters. In Europe, negotiated agreements between government and individual companies or industry sectors have rapidly become one of the principal environmental management and policy instruments at a national level. The goal here is often to fill in the gaps not covered by regulations, to encourage companies to go beyond compliance, or even to find a more politically acceptable alternative to regulation. Public voluntary programmes such as those described in Chapter 6 also fit the partnership model, with government offering technical support and public relations benefits in return for industry commitments to improved environmental performance. In the United States they have also become an important component of 'Reinventing Environmental Regulation' (Clinton and Gore Jr 1995), under which government seeks to replace the typically adversarial relationship that has existed between business and government in that country with a more co-operative approach based on trust and reciprocity.

A variety of other kinds of environmental partnership were considered in our case studies. In the case of the Victorian vegetable-growing industry, a limited industry–VEPA partnership had the potential to achieve far greater environmental gains by incorporating the supply chain and the community. In the case of vehicle smash repairs, a cleaner-production partnership between VEPA and the industry promised to harness significant win–win opportunities. And, in the case of Environmental Improvement Plans and accredited licensing, a tripartite partnership between VEPA, community and business aspired both to achieve a more co-operative relationship between government and industry and to directly involve the community in negotiations and oversight as part of the broader enterprise of facilitative regulation.

Civil regulation and participatory governance

As defined by Bendell (1998: 8): 'civil regulation is where organisations of civil society,[8] such as NGOs, set the standards for business behaviour. Enterprises then choose to adopt or not to adopt those standards.' Those who advocate a greater role for civil regulation

---

8   Civil society is conventionally defined as involving 'citizens acting collectively in a public sphere to express their interests, passions, and ideas, exchange information, achieve mutual goals, make demands on the state, and hold public officials accountable. Civil society is an intermediary entity, standing between the private sphere and the state' (Diamond 1996).

argue that the regulatory state is starved of resources, lacking in political will, and incapable of reaching the many businesses that can now operate outside national territorial boundaries. The goal of civil regulation[9] is to fill the vacuum left by the contracting state, and to compensate for 'the deficit of democratic governance that we face as a result of economic globalisation' (Bendell 2000: 201). As such, there is considerable overlap between this perspective and some aspects of regulatory pluralism, discussed above.

Under civil regulation, the various manifestations of civil society act in a variety of ways to influence corporations, consumers and markets, often bypassing the state and rejecting political lobbying in favour of what they believe to be far more effective strategies. Sometimes NGOs take direct action, usually targeted at large reputation-sensitive companies. Greenpeace's campaign against Shell's attempted deep-sea disposal of the Brent Spar oil rig is one example. Sometimes, they boycott products or producers deemed to be environmentally harmful, as with the effective boycott of Norwegian fish products organised by Greenpeace in protest against that nation's resumption of whaling. Market campaigning, focusing on highly visible branded retailers, is a particularly favoured strategy.[10] Less so are campaigns that seek to provide a market premium for 'environmentally preferred' produce, due largely to the unwillingness of consumers to support such a strategy. More recently, certification programmes such as the Forest Stewardship Council are 'transforming traditional power relationships in the global arena. Linking together diverse and often antagonistic actors from the local, national and international levels . . . to govern firm behavior in a global space that has eluded the control of states and international organizations' (Gereffi et al. 2001; see also Gamble and Ku 2000).

However, the evolving role of civil regulation has not taken place entirely divorced from state intervention. On the contrary, either in response to pressure from the institutions of civil society or in recognition of the limits of state regulation, governments are gradually providing greater roles for communities, environmental NGOs and the public more generally. Thus a number of next-generation policy instruments examined in the previous chapters are geared to empower various institutions of civil society to play a more effective role in shaping business behaviour. In effect, they facilitate civil regulation (and regulatory pluralism). These include public participation provisions under the various United States Reinventing Environmental Regulation initiatives, CRTK legislation, some second-generation voluntary agreements which contemplate a significant role for third parties, and some forms of environmental partnership in which the public, or public-interest groups, are major players.

Arguably, the most powerful forms of civil regulation are those in which environmental NGOs or communities have the capacity to threaten the social licence and reputation capital of large corporations. Sometimes they do so independent of government, but more commonly government, and next-generation instruments, play a crucial facilitative role. We provided a number of examples of how this might come about in the preceding chapters. These include the contribution of CLCs in EIPs, as considered in Chapter 8, the

9   The term is used as a variant of civil society, not to imply the use of civil rather than criminal law.

10   For example, a highly successful Greenpeace campaign has been largely responsible for sensitising European consumers (particularly in Germany and the UK) to the clear-felling of old-growth forests. This has had a profound impact on North American companies exporting to those markets, as they are increasingly being pressured by European buyers to provide evidence that the timber they supply has come from sustainably harvested sources.

capacity for third-party oversight and shaming to provide an effective underpinning to voluntary initiatives, as with the mining industry examined in Chapter 7, and the highly effective model of the Indonesian PROPER PROKASIH programme, described in Chapter 6, which also demonstrated how government and civil society can work closely together to achieve the best results.

Ecological modernisation and the 'greengold thesis'

Another emerging paradigm is what has become known as ecological modernisation.[11] In contrast to many analyses which suggest that a radical reorientation of our current economic and social arrangements will be necessary to avert ecological disaster (see e.g. Commoner 1972; O'Connor 1996), ecological modernisation suggests that ecologically sound capitalism is not only possible but worth working towards.[12] This good-news message may indeed be a substantial part of the attraction of the ecological modernisation approach. Beyond this, the main tenets of this perspective are difficult to encapsulate, since writings under the ecological modernisation banner are diverse and draw from a number of different schools of thought.[13]

For present purposes we focus on its core, which emphasises how strategies such as eco-efficiency can facilitate environmental improvements in the private sector (particularly in relation to manufacturing) by simultaneously increasing efficiency and minimising pollution and waste (Buttel 2000: 59). This will require switching to the use of cleaner, more efficient and less resource-intensive technologies, shifting away from energy and resource-intensive industries to those that are value- and knowledge-intensive, anticipatory planning processes, and the 'organisational internalisation of ecological responsibility' (Cohen 1997: 109).

However, this is not to suggest that markets unaided, or past environmental policy, will provide the appropriate messages and incentives to enable industry to achieve these goals. On the contrary, such an outcome requires action on a number of fronts, and government regulation in particular will need to promote innovation in environmental technology. In terms of public policy prescriptions, Mol (one of the most influential proponents of this perspective) suggests two directions that should be pursued. First, state environmental policy must focus not on prescription but rather on prevention and participatory decentralised decision-making, which 'creates favourable conditions and contexts for environmentally sound practices and behaviour on the part of producers and consumers' (Mol 1995: 46). The second option includes a transfer of responsibilities, incentives and tasks from the state to the market, which provides the flexibility and incentives to enable more efficient and effective outcomes. Under this approach 'the state provides the conditions and stimulates social "self-regulation", either via economic mechanisms and dynamics or via the public sphere of citizen groups, environmental NGOs and consumer organisations' (Mol 1995: 47).

In these respects, the ecological modernisation literature has resonance with a number of other perspectives described in this chapter, especially civil society, regulatory plural-

---

11 For an overview of the literature on ecological modernisation, see Mol and Sonnenfeld 2000. See also *Futures* 2000.
12 For a more nuanced statement of this position, see Mol and Sonnenfeld 2000.
13 For an excellent overview, see Buttel 2000.

ism and, to some extent, reflexive regulation. However, on one fundamental issue eco-logical modernisation departs substantially from these other perspectives: namely, in its assumption that by following the precepts of ecological modernisation there will be a 'dissolution of the conflict between economic progress and responsible environmental management because it will be possible to achieve both objectives simultaneously' (Cohen 1997: 109).

In arguing that the business community could successfully combine the objectives of environmental protection and economic growth, ecological modernisation resonates with the views of a variety of business strategists, environmental commentators and corporations who subscribe to what has become known as the 'greengold thesis'. This group argues that, by preventing pollution and thereby cutting costs and avoiding waste directly, by more effective risk management, by gaining an increasing share of expanding 'green markets' or price premiums within them, and by developing the environmental technology to compete effectively in the global environmental market, businesses can achieve win–win outcomes, gaining economically from environmental improvements (Smart 1992; Schmidheiny 1992).

Of particular influence have been the views of Porter (1991), who has argued that, in a highly regulated world, innovative companies can acquire competitive advantages or cut costs by developing novel methods of reducing environmental problems. Notwithstand-ing some differences of emphasis, a common refrain has been that going beyond compliance was both good for business and good for the environment (see generally Gunningham 1994). However, both Porter and the ecological modernisation theorists acknowledge that there may be more scope for win–win outcomes in some sectors and circumstances than in others (Porter 1998; Baylis et al. 1998).

A number of next-generation instruments might facilitate win–win outcomes. For example, instruments that harness market forces, so as to encourage rather than inhibit commercial drive and innovation (including many economic instruments and perfor-mance standards)[14] meet with approval. Various other flexible and arguably cost-efficient mechanisms for curbing environmental degradation, such as self-regulation, informa-tion-based strategies, the use of liability rules and other financial instruments, are consistent with Mol's two directions summarised above. Government's role includes nudging firms towards cleaner production, heightening their awareness of environ-mental issues, providing them with financial incentives (which at the margin may be crucial), and encouraging the re-ordering of corporate priorities in order to reap the benefits of improved environmental performance.

The question of whether in a particular set of circumstances there are opportunities for win–win outcomes or not is both highly contentious and important, because, in the absence of such opportunities, it cannot be assumed that organisations will voluntarily become greener, or that they have any incentive to pursue beyond-compliance environ-mental strategies in the absence of external pressure to do so. As regards the latter issue, Reinhardt (2000) has demonstrated that it makes sense to pursue beyond-compliance

---

14   For example, Porter and van der Linde's (1995) argument is that a well-designed regulatory system can foster innovation by concentrating on outcomes (i.e. performance standards) rather than on techniques (i.e. specification- or technology-based standards). This is because performance standards free an enterprise to respond to a regulator's requirement in the way it thinks fit.

policies if they increase the enterprise's expected value, or if they appropriately manage business risk, but in a substantial number of circumstances they do neither.

Regulatory reform: the never-ending journey

While each of the perspectives described above provides insights concerning how best to approach the task of regulatory reconfiguration, there are considerable disparities between them, and none provides unproblematic or comprehensive answers as to what next-generation environmental regulation should involve. Nevertheless, both the commonalities and the differences between these perspectives provide insights as to how best to approach the journey ahead.

In terms of the former, there is general agreement that returning to the policies of the past is not an option. A common theme is that traditional regulation is not suited to meet many contemporary policy needs (although, as we emphasise below, it still has a role to play), and indeed it is partly in response to the perceived shortcomings of the regulatory status quo that each of these conceptual frameworks evolved. As Fiorino (1999: 464) puts it: 'underlying each strand in the literature is the belief that the increased complexity, dynamism, diversity, and interdependence of contemporary society makes old policy technologies and patterns of governance obsolete'.

There is also recognition that regulated enterprises have a diversity of motivations and that it cannot be assumed (as in some versions of command-and-control regulation) that deterrence is the principal weapon available to regulators and policy-makers. Notwithstanding differences of emphasis, there is a shared awareness of the complexity of motivational forces influencing environmental behaviour, and of the need to develop instruments and strategies to take account of this. As a result, each of these perspectives to a greater or lesser extent recognises the importance of such broader motivational drivers as the effects of negative publicity, informal sanctions and shaming, incentives provided by various third parties, the significance for private enterprise of maintaining legitimacy, and the necessity to maintain co-operation and trust.[15]

Again, some instruments and approaches are common to almost all of these perspectives. For example, informational regulation is important to reflexive regulation, civil regulation and regulatory pluralism; is supportive of environmental partnerships; and is at least consistent with the goals of ecological modernisation. Similarly, process-based strategies such as EMS are central to reflexive regulation, many environmental partnerships and ecological modernisation and, as a form of industry self-management, to some variants of regulatory pluralism. And none of these perspectives would deny that there is a role for public-interest groups, although their role is conceived as more central in the case of environmental partnerships, regulatory pluralism and civil regulation than it is for reflexive regulation or ecological modernisation.

And, in contrast to traditional forms of environmental regulation, each of the perspectives we have examined sees virtue in engaging with environmental leaders and in encouraging or rewarding their further improvement rather than focusing only on bring-

---

15   For an excellent summary of the recent literature on these issues, see Parker 2000.

ing laggards up to compliance. This is perhaps most obvious in the case of environmental partnerships (which often only the best firms are willing to join), but regulatory pluralism and civil regulation also reward environmental leaders (for example, in terms of reputation, or market advantage, or share price premium), as well as seeking to shame or otherwise provide negative incentives to laggards. Reflexive regulation, while less explicit, builds in processes that often lead to continuous improvement, while ecological modernisation, with its emphasis on win–win outcomes and cleaner production, also seeks to encourage best practice rather than merely minimum standards and compliance.

However, when it comes to identifying where the focus of regulatory reconfiguration should be, there is much less agreement, and very different policy prescriptions flow from different perspectives. In terms of reflexive regulation, the perceived solution is to establish regulatory structures that strengthen the capability of individual institutions or enterprises for internal reflection and self-control. For regulatory pluralism, it is a plethora of instruments that enable the state to 'steer not row', and to harness the capacities of second and third parties to more effectively fill the space vacated by the contracting regulatory state. From a civil regulation perspective, the state's principal role is to provide mechanisms that will empower the institutions of civil society to make corporations more accountable. A partnership perspective would seek out opportunities to build reciprocal gains from co-operation, with the state playing an additional role as facilitator. For ecological modernisation, the aspiration is to create incentives that will help industry move towards sustainability using new technologies and techniques of production, with economic and environmental considerations being mutually reinforcing.

Our own perspective is that each of the above frameworks has something valuable to offer and that none of them is 'right' or 'wrong' in the abstract. Rather, they make differing contributions depending on the nature and context of the environmental policy issue to be addressed. In the preceding chapters, we provided many examples of why different policies work in different circumstances Indeed, the dichotomy between small and medium-sized enterprises and large companies, which formed the basis for the structure of the book, is one illustration of this. It is a small step to make a similar point couched in terms of the conceptual frameworks examined immediately above, and it should be readily apparent why different conceptual frameworks have more or less resonance in different contexts.

For example, ecological modernisation has most to offer where industry can demonstrably benefit economically from environmental improvements (the so-called win–win scenario),[16] but is far less persuasive in the variety of contexts where this is not the case.[17] Civil regulation has considerable power when it comes to changing the behaviour of large reputation-sensitive enterprises, which are vulnerable not only to shaming but also to

16   For example, there may be considerable win–win opportunities in relation to SMEs (there is much low-hanging fruit to be taken advantage of) and an important contribution can be made by mechanisms that seek out cleaner production and eco-efficiency opportunities.

17   The evidence suggests that the range of circumstances in which it is possible for enterprises to benefit both the environment and their own economic bottom line (even with certain policy interventions) is more limited than many ecological modernisation theorists and their fellow travellers have claimed. Many of the best essays on this debate, couched mainly in terms of Porter's greengold thesis, are contained in Harvard Business Review on Business and the Environment 2000. See also Newton and Harte 1997; Howes *et al.* 1997: 86; Reinhardt 2000.

market forces and consumer pressure, but has far less to offer when dealing with the environmental excesses of many SMEs and firms that are not vulnerable to such pressures. Reflexive regulation is demonstrably effective in dealing with complex and sophisticated environmental issues such as regulating major hazardous facilities,[18] but may be redundant when it comes to more traditional challenges.[19] Environmental partnerships have attractions where both partners can see common ground and mutual benefits from constructive engagement,[20] but not where there are irreconcilable philosophical differences between stakeholders (Poncelet 1999).

The limitations of each of the major policy innovations, and of the conceptual frameworks that drive next-generation regulation, lead to a plea for pragmatism and regulatory pluralism. None of the policy instruments or perspectives we have examined works well in relation to all sectors, contexts or enterprise types.[21] Each has weaknesses as well as strengths, and none can be applied as an effective stand-alone approach across the environmental spectrum. In part, such a conclusion suggests the value of designing complementary combinations of instruments, compensating for the weaknesses of each with the strengths of others, while avoiding combinations of instruments deemed to be counterproductive or at least duplicative. This indeed was the central message of our previous work, embedded within the pluralist perspective (Gunningham and Sinclair 1999a, 1999b). From this perspective, no particular instrument or approach is privileged, whether it is reflexive regulation, civil regulation or the tenets of ecological modernisation. Rather, the goal is to accomplish substantive compliance with regulatory goals by any viable means using whatever regulatory or quasi-regulatory tools might be available, including any or all of the next-generation instruments described in preceding chapters. As Parker (2000) points out: 'the objective is to steer corporate conduct towards public policy objectives in the most effective and efficient way, without interfering too greatly with corporate autonomy and profit, rather than fruitless expenditure of government and business resources on traditional styles of regulation that ignore the effects of indigenous regulatory orderings'.

18 Such facilities are extremely complex, and demonstrably beyond the reach of rule-based regulation. Policy-makers have relied instead on the use of risk and environmental management systems, and internal self-control, with apparent success. After the Three Mile Island near-meltdown, the nuclear safety regime in the United States shifted from rule-based regulation to a new paradigm, which Braithwaite and Drahos (2000) characterise as regulatory scrutiny of risk management systems and shaming within the nuclear professional community of companies that failed to improve their systems. Within a decade SCRAMS (safety-related automatic shutdowns of nuclear plants) fell sevenfold. See also Rees 1994.

19 For example, the most effective means of dealing with highly hazardous pesticides may be simply to ban their use. Again, reflexive regulation can only play a minor role in curbing the environmental excesses of small and medium-sized enterprises, perhaps in terms of simplified EMSs.

20 As we indicated in our case studies, environmental partnerships are much more likely to flourish in situations where there is a genuine commonality of interest between the environment and economic interests, and, where this does not exist, may only be credible when underpinned by the threat of direct intervention if the partnership fails to achieve specified public-interest objectives. The experience of voluntary agreements in Europe, described in Chapter 6, would also support this view.

21 Indeed, apart from regulatory pluralism (which almost by definition is catholic in its approach), each has only a limited sphere in which it is likely to be effective.

However, even in circumstances where one particular perspective (or combination of perspectives) and one set of policy tools seem well suited to a particular problem, there may still be a substantial gap between theory and practice. Indeed, some of the policies at the very heart of next-generation regulation are largely untested and their efficacy is uncertain. This is certainly the case with environmental management systems, which play an important role under a number of the frameworks we reviewed above. There is only very limited evidence available of how they work in practice (mainly in relation to major hazard facilities) and, as our overview in Chapter 6 and our case study in Chapter 8 demonstrated, there remains a risk that they will produce the trappings of self-reflection and internal control without achieving more than 'business as usual'. Moreover, it has proved very difficult to develop incentives sufficient to persuade substantial numbers of organisations to participate in an EMS-based alternative regulatory track. And we know even less about whether or to what extent a 'slimmed-down' version of this approach might be applied to SMEs.

Similarly, our examination of voluntary agreements in Europe suggested that the first generation of these agreements had had very substantial deficiencies. Some second-generation agreements are much better designed but we still have incomplete evidence as to whether, or under what circumstances, they will be successful. Indeed, there remains a considerable risk of wrong turnings and of re-enacting the mistakes of previous decades.[22] Much the same can be said for many environmental partnerships (Gunningham and Sinclair 2002). Even informational regulation, which has generally been hailed as a success story, has been challenged by its critics for not demonstrably achieving many of its objectives, at least in some jurisdictions (Antweiler and Harrison 1999). The limitations of our current experience are even greater in the case of SMEs, where, as we saw in Chapter 2, the empirical picture remains extremely unclear.

Thus much of our knowledge about policy instruments, and in particular about what works and when, is tentative, contingent and uncertain. This suggests the virtue of adaptive learning, and for treating policies as experiments from which we can learn and which in turn can help shape the next generation of instruments. From this perspective, following Fiorino (1999: 468), it is important to ask: 'how may mechanisms that promote policy-learning . . . be strengthened? To what extent do policy-making institutions provide mechanisms for learning from experience and altering behavior based on that experience?' This might imply, for example, monitoring, ex post evaluation and revision mechanisms and

> building reliable feedback mechanisms into policy-making, strengthening learning networks, creating conditions that would lead to more trust and more productive dialogue and building enough flexibility into the policy system so that it is possible to respond to lessons drawn from one's own experience or that of others (Fiorino 1999: 468).

In particular, adaptive learning is heavily dependent on the depth and accuracy of an agency's statistical database and other information sources. Only with adequate data collection and interpretation can one know how effective or otherwise a particular regu-

---

22 For example, it has been pointed out that voluntary approaches, 'while enjoying a resurgence . . . in fact represent the dominant approach throughout history. Self-regulation was the norm prior to the 1970s, and its failure was the reason we started regulating in the first place' (Andrews 1998: 179, quoted in Harrison 2001).

latory strategy has been. There will be a need to establish databases that provide more accurate profiles of individual firms, hazards and industries.[23] Environmental information systems have the potential to play a key role here. Work in this area is still in its embryonic stage, but recent initiatives (Rapp *et al.* 2000) suggest it is developing quite rapidly. Of particular note is the Finnish compliance monitoring system (VAHTI), which comprises a database for the input and storage of information on the environmental permits of industry and their discharges into water, emission to air and solid wastes.[24]

Finally, we return to the role of direct state regulation under a next-generation approach. We do so because it is important not to lose sight of the residual but nevertheless important role that command-and-control regulation can and should continue to play in environmental policy. It is only the state that can impose criminal sanctions and the full weight of the law, and only the state that, under statute, may have power of entry into private property to inspect, take samples and gather evidence of illegality more generally. While there may be some circumstances where—as advocates of civil regulation, reflexive regulation and regulatory pluralism would argue—far more can be achieved by various other forms of state and non-state action, this is certainly not the case across the board.

For example, there remain situations where SMEs in particular need the highly specific and concrete guidance that specification standards can provide. And, in the case of large companies, the most important 'step' changes in environmental performance in industries such as pulp and paper have been achieved through mandated technological change (Gunningham *et al.* forthcoming). It should also be remembered that, according to various surveys, the single most important motivator of improved environmental performance is regulation (Henriques and Sadorsky 1995; ENDS 1997). The more general conclusion, as the United States EPA (2000a: 4) has recognised, is that: 'in some cases, nationwide laws and regulations will continue to be the best way to reduce risk. But in others, tailored strategies that involve market based approaches, partnerships, or performance incentives may offer better results at lower costs.'

The broader point is that many less interventionist strategies are unlikely to succeed if they are not underpinned by direct regulation. For example, under reflexive regulation, some enterprises may be tempted to develop 'paper systems' and tokenistic responses which 'independent' third-party auditors may fail to detect (O'Rourke 2000). However, the threat of sanctions if they fail to deliver on performance targets set by the state will substantially reduce the risk of free-riding. Similarly, in Chapter 4 we demonstrated how a regulatory oversight role was crucial to the success of an industry self-management

23 Note the US EPA's Office of Enforcement and Compliance Assurance National Performance Measures Strategy, which is developing a measurement framework of outputs, outcomes and environmental indicators. See http://es.epa.gov/oeca/perfmeas/npmsfinal.html.

24 The main value of the database is to act as a tool for regulators, to help them estimate the level of compliance monitoring and focus their resources on areas where they are most needed, and thereby to facilitate better processing and monitoring of permits. In future the database will be accessible to the public and as such will also facilitate CRTK (see Chapter 6). In the European Union, the value of such systems is likely to be increased by a number of current initiatives. In particular, the European Commission has determined that member states must identify installations and discharges falling within the scope of the Integrated Pollution Prevention and Control (IPPC) Directive and collect information on emissions and discharges from those installations and facilities and report to the Commission necessary information. See in particular Hietamaki 2000a, 2000b; Koski 2000.

initiative in the refrigeration and air-conditioning industry, and that co-regulation, not self-regulation, provides the best hope of success. Again, there is evidence that information-based strategies cannot necessarily replace traditional regulation and enforcement practices but rather that the two instruments work best when used in a complementary combination (Foulon *et al.* 1999). We also showed how, in the case of small business, the fear of regulation or its enforcement can be used to good effect to complement other more innovative approaches.

Once again, what we are witnessing is not the demise of the regulatory state but a regulatory reconfiguration, in which command-and-control retains a place, albeit no longer at centre stage but rather as a complement to a range of next-generation policies. But this reconfiguration remains a work in progress. Certainly, our knowledge of what works and why is much greater than it was a decade ago, and both the central chapters and case studies in this book have demonstrated impressive advances over the previous generation of policy instruments. Nevertheless, our research equally demonstrates that the journey to best-practice environmental regulation is far from complete. Notwithstanding the considerable promise of the new generation of environmental policy tools, the road to regulatory reform is long and tortuous, and the journey is far from over.

BIBLIOGRAPHY

ABS (Australian Bureau of Statistics) (2000) 'Industry Sub-sectors' (unpublished).

Adams, J. (1999) 'Foreign Direct Investment and the Environment: The Role of Voluntary Corporate Environmental Management', in *OECD Proceedings: Foreign Direct Investment and the Environment* (Paris: OECD).

Afsah, S., and T. Vincent (1997) 'Putting Pressure on Polluters: Indonesia's PROPER Program' (Harvard Institute for International Development, www.worldbank.org/nipr/work_paper/vincent).

Afsah, S., A. Blackman and D. Ratunanda (2000) 'How do Public Disclosure Pollution Control Programs Work? Evidence from Indonesia' (Resources for the Future Discussion Paper 00–44; www.rff.org/CFDOCS/disc_papers/pdf_files/0044.pdf, October 2000).

AIIIC (Automotive Industry's International Information Center) (2000) 'News Events' (www.ciia.com/provinces/ontario/newsevents.html).

Andrews, R.N.L. (1998) 'Environmental Regulation and Business "Self-regulation" ', *Policy Sciences* 31: 177-97.

Angarwal, A. (1999) 'Enter the Green Rating Project', *Down to Earth* 8.5 (31 July 1999): 1-5.

Anleu, S.L., L. Mazerolle and L. Presser (2000) 'Third Party Policing and Insurance', *Law and Policy* 22.1: 67-87.

Antweiler, W., and K. Harrison (1999) 'Environmental Regulation vs. Environmental Information: A View from Canada's National Pollutant Release Inventory' (paper presented to the Association for Public Policy Analysis and Management, Washington, DC, November 1999).

APEC (Asia Pacific Economic Co-operation) (1999) *Eco-Efficiency in Small and Medium Enterprises* (final report; www.actetsme.org/archive/eco-efficiency/eco-efficiency.html).

Arora, S., and T.N. Cason (1995) 'An Experiment in Voluntary Environmental Regulation: Participation in EPA's 33/50 Program', *Journal of Environmental Economics and Management* 28.3: 271.

Ayres, I., and J. Braithwaite (1992) *Responsive Regulation* (Oxford, UK: Oxford University Press).

Baeke, S., M. De Clercq and F. Matthijs (1999) 'The Nature of Voluntary Approaches: Empirical Evidence and Patterns' (CAVA [Concerted Action on Voluntary Approaches] Working Paper, 99/08/3, www.cerna.ensmp.fr/Documents/99-08-3.pdf).

Bates, G. (1995) *Environmental Law in Australia* (Sydney: Butterworths, 4th edn).

Baylis, R., L. Cornnell and A. Flynn (1998) 'Sector Variation and Ecological Modernization: Towards an Analysis at the Level of the Firm', *Business Strategy and the Environment* 7: 150-61.

Baumol, W., and W. Oates (1998) *The Theory of Environmental Policy* (New York: Cambridge University Press).

Beardsley, D. (1996) *Incentives for Environmental Improvement: An Assessment of Selected Innovative Programs in the States and Europe* (Washington, DC: Global Environmental Management Initiative).

Bendell, J. (1998) 'Editorial', *Greener Management International* 24 (Winter 1998): 3-9.

—— (2000) 'Civil Regulation: A New Form of Democratic Governance for the Global Economy', in J. Bendell (ed.), *Terms of Endearment: Business, NGOs and Sustainable Development* (Sheffield, UK: Greenleaf Publishing): 239-54.

Bickerdyke, I., and R. Lattimore (1997) *Reducing the Regulatory Burden: Does Firm Size Matter?* (Melbourne: Industry Commission).

Bomsel, O., P. Borkey, M. Glachant and F. Lévêque (1996) 'Is There Room for Environmental Self-regulation in the Mining Sector?', Resources Policy 22.2: 79-86.

Borkey, P., and F. Lévêque (1998) Voluntary Approaches for Environmental Protection in the European Union (OECD Working Paper, ENV/EPOC/Gee1 [98]29: 5; Paris: Organisation for Economic Co-operation and Development).

Boyd, J., A. Krupnick and J. Mazurek (1998) Intel's XK Permit: A Framework for Evaluation (Resources for the Future Discussion Paper 98-11; Santa Barbara, CA: School of Environmental Science and Management).

Boyle, G. (1999) 'Refrigerants Seminar', News and Views (AIRAH WA Division) 6: 3.

Braithwaite, J. (1989) Crime, Shame and Integration (Cambridge, UK: Cambridge University Press).

—— (1993) 'Responsive Business Institutions', in C. Cody and P. Sampford (eds.), Business, Ethics and Law (Sydney: Federation Press).

—— (1999) 'Restorative Justice: Assessing an Immodest Theory and a Pessimistic Theory', Crime and Justice 25: 448.

Braithwaite, J., and P. Drahos (2000) Global Business Regulation (Cambridge, UK: Cambridge University Press).

Briody, M., and T. Prenzler (1998) 'The Enforcement of Environmental Protection Laws in Queensland: A Case of Regulatory Capture?', Environmental and Planning Law Journal 15: 54.

Brown, R. (1992) 'Administrative and Criminal Penalties in the Enforcement of Occupational Health and Safety Legislation', Osgoode Hall Law Journal 30.3: 691.

Burtraw, D. (1998) Cost Savings, Market Performance, and Economic Benefits of the US Acid Rain Program (Discussion Paper 98-28REV; Washington, DC: Resources for the Future).

Burtraw, D., A.J. Krupnick, E. Mansur, D. Austin and D. Farrell (1997) The Costs and Benefits of Reducing Acid Rain (Discussion Paper 97-31REV; Washington, DC: Resources for the Future).

Buttel, F.H. (2000) 'Ecological Modernization as Social Theory', Geoforum 31: 57-65.

CEC (Commission of the European Communities) (1996) Communication from the Commission to the Council and the European Parliament on Environmental Agreements, COM(96)561, Brussels.

Center for Strategic and International Studies (1998) The Environmental Protection System in Transition: Towards a More Desirable Future (Final Report of Enterprise for the Environment; Washington, DC: Center for Strategic and International Studies).

Chertow, M., and D. Esty (eds.) (1997) Thinking Ecologically: The Next Generation of Environmental Policy (New York: Yale University Press).

Clay, S. (1998) Environment Management Experts Identify Obstacles to ISO 14001 (INEM [International Network for Environmental Management], www.inem.org/htdocs/iso/iso-sme.html).

Clinton, W.J., and A. Gore, Jr (1995) Reinventing Environmental Regulation (Washington, DC: Environmental Protection Agency).

Coglianese, C., and J. Nash (2001) Regulating from the Inside: Can Environmental Management Systems Achieve Policy Goals? (Washington, DC: Resources for the Future).

Cohen, D.S. (1991) 'The Regulation of Green Advertising: The State, the Market and the Environmental Good', University of British Columbia Law Review 25: 225.

—— (2002) 'Voluntary Codes: The Role of the State in a Privatized Regulatory Environment, in K. Webb (ed.), Voluntary Codes: Private Governance, the Public Interest and Innovation (Ottawa: Carleton University, Research Unit for Innovation, Science and the Environment).

Cohen, M.J. (1997) 'Risk Society and Ecological Modernisation', Futures 29.2: 105-19.

Cohen, S. (1986) 'EPA: A Qualified Success', in S. Kamieniecki, R. O'Brien and M. Clarke (eds.), Controversies in Environmental Policy (Albany, NY: SUNY Press): 174.

Cole, D.H., and P. Grossman (1999) 'When is Command and Control Efficient? Institutions, Technology and the Comparative Efficiency of Alternative Regulatory Regimes for Environmental Protection', Wisconsin Law Review 1999/5: 887-939.

Commission for Environmental Co-operation (1998) Voluntary Measures to Ensure Environmental Compliance (Montreal: Commission for Environmental Co-operation).

Commoner, B. (1972) The Closing Circle: Nature, Man and Technology (New York: Bantam Books).

Covery, F., and F. Lévêque (2001) 'Applying Voluntary Approaches: Some Insights from Research', presented at CAVA (Concerned Action on Voluntary Approaches) International Policy Workshop on the Use of Voluntary Approaches, Brussels, 1 February 2001, www.cerna.ensmp.fr/Progeuropeens/CAVA/Workshops.html.

Crow, M. (2000) 'Beyond Experiments', *The Environmental Forum*, May/June 2000: 19-29.

Cullen, Lord (chairman) (1990) *Piper Alpha Inquiry* (London: HMSO).

Davies, T., and J. Mazurek (1996) *Industry incentives for Environmental Improvement: Evaluation of US Federal Initiatives* (Washington, DC: Global Environment Management Initiative).

Dawson, S., and N. Gunningham (1996) 'The more dolphins there are the less I trust what they're saying: Can green labelling work?', *Adelaide Law Review* 18.1: 1-34.

De Bruijn, T., and K. Lulofs (2000) 'Driving Small and Medium-Sized Enterprises towards Environmental Management: Policy Implementation in Networks', in R. Hillary (ed.), *Small and Medium-Sized Enterprises and the Environment: Business Imperatives* (Sheffield, UK: Greenleaf Publishing): 263-74.

De Clercq, M., A. Seyad, A. Suck and B. Ameels (undated) *A Comparative Study of Environmental Negotiated Agreements* (NEAPOL, http://fetew.rug.ac.be/neapol/conference/index.html).

DEP (Western Australian Department of Environmental Protection) (1998) *Best Practice Environmental Licences* (Perth, Australia: DEP).

Diamond, L. (1996) 'Towards Democratic Consolidation', in L. Diamond and M. Platter (eds.), *The Global Resurgence of Democracy* (Baltimore, MD: The Johns Hopkins University Press): 228.

Eakin, J. (1992) 'Leaving it up to the Workers: A Sociological Perspective on the Management of Health and Safety in Small Workplaces', *International Journal of Health Services* 22: 698-704.

Eckersley, R. (1995) *Markets, the State and the Environment* (Melbourne: Macmillan Press).

ECOTEC Research and Consulting (2000) *Report on SMEs and the Environment* (Brussels: European Commission, Directorate General Environment, www.europa.eu.int/comm/environment/sme/index.htm).

EEA (European Environment Agency) (1997) *Environmental Agreements: Environmental Effectiveness* (Environmental Issues Series, 3.1–2; Copenhagen: EEA).

ENDS (Environmental Data Services) (1997) 'DoE Rediscovered Business Benefits of Environmental Policy', *ENDS Report* 266 (March 1997): 24.

—— (1999a) 'Setting the Agenda for the Revision', *ENDS Report* 289 (February 1999): 12-13.

—— (1999b) 'New Environmental Management Models for SMEs', *ENDS Report* 290 (March 1999): 5.

—— (1999c) 'DTI Study Identifies Barriers to Small Firms' Uptake of EMS', *ENDS Report* 299 (December 1999): 9.

—— (2000a) 'Revised EU Ecolabel Regulation Approved', *ENDS Environment Daily*, 3 July 2000.

—— (2000b) 'EU Ecolabel Scheme "Turning the Corner"', *ENDS Environment Daily*, 1 December 2000.

—— (2000c) 'Norway Puts Firms' Green Records on the Web', *ENDS Environment Daily*, 6 December 2000.

—— (2000d) 'How Effective are Environment Management Systems?' *ENDS Report* (December 2000): 311.

Environmental Forum (1999) 'Another Parallel Track Proposal?', *The Environmental Forum*, November/ December 1999: 47-54.

Environmental Manager (2000) *Environmental Manager* 315 (4 October 2000).

EPA (Western Australian Environment Protection Authority) (1993) *New Ozone Protection Law for Western Australia 1993* (Perth, Australia: EPA).

Fanshawe, T. (2000) 'The Interrelationship between Environmental Regulators, Small and Medium-Sized Enterprises and Environmental Help Organisations', in R. Hillary (ed.), *Small and Medium-Sized Enterprises and the Environment: Business Imperatives* (Sheffield, UK: Greenleaf Publishing): 244-53.

Fiorino, D. (1999) 'Rethinking Environmental Regulation: Perspectives from Law and Governance', *Harvard Environmental Law Review* 23.2: 441-69.

Fischer, K., and J. Schot (eds.) (1993) *Environmental Strategies for Industry* (Washington, DC: Island Press).

Foulon, J., P. Lanoie and B. Laplante (1999) *Incentives for Pollution Control: Regulation and Public Disclosure* (Policy Research Working Paper; Washington, DC: World Bank).

Freeman, J. (2000) 'The Contracting State', *Florida State University Law Review* 28.1 (Fall 2000).

FOE (Friends of the Earth) (2000) 'Insurance Firms Named and Shamed: Record Feeble on Ethical Investment', www.foe.co.uk/pubsinfo/infoteam/pressrel/2000/20000119000148.html.

Fung, A., and D. O'Rourke (2000) 'Reinventing Environmental Regulation from the Grassroots Up: Explaining and Expanding the Success of the Toxics Release Inventory', *Environmental Management* 25.2 (February 2000): 115-27.

Furger, F. (2000) 'Voluntary Codes between Effectiveness and Legitimacy', presented at a workshop on *Education, Information and Voluntary Measures in Environmental Protection*, National Academy of Sciences/ National Research Council Committee on the Human Dimensions of Global Change, Washington, DC, November 2000.

Futures (2000) *Futures* 33.3–4 (April/May 2000; Special Issue on 'Ecological Modernization: Citizenship and Ecological Modernization in the Information Society').

Gamble, J.K., and C. Ku (2000) 'International Law: New Actors and New Technologies: Center Stage for NGOs?', *Law and Policy in International Business* 31: 221-62.

Guardian (2001) 'Green tax scheme flawed, say auditors', *The Guardian* 9 February 2001: 4.

GEMI (Global Environmental Management Initiative) (1992) *Total Quality Environmental Management* (Washington, DC: GEMI).

—— (1999) *Environmental Improvements through Business Incentives* (www.gemi.org/IDE_004.pdf).

Gereffi, G., R. Garcia-Johnson and E. Sass (2001) 'The NGO–Industrial Complex', *Foreign Policy*, July/August 1999 (www.foreignpolicy.com/issue_julyaug_2001/gereffi.html).

Gerstenfeld, A., and H. Roberts 'Size Matters: Barriers and Prospects for Environmental Management in Small and Medium-Sized Enterprises', in R. Hillary (ed.), *Small and Medium-Sized Enterprises and the Environment: Business Imperatives* (Sheffield, UK: Greenleaf Publishing): 106-18.

Gibson, R. (2000) *Voluntary Approaches: The New Politics of Corporate Greening* (Ontario: Broadview Press).

Gould, I. (2000) 'The Code: Driving Change', *Groundwork* 7.4 (September 2000): 1-4.

Gouldson, A., and J. Murphy (1998) *Regulatory Realities* (London: Earthscan Publications).

Grabosky, P. (1995) 'Using Non-government Resources to Foster Regulatory Compliance', *Law and Policy* 17.3: 256-81.

Gray, W.B., and J.T. Scholz (1993) 'Does Regulatory Enforcement Work? A Panel Analysis of OHSA Enforcement', *Law and Society Review* 27.1: 177-215.

Grodsky, J.A. (1993) 'Certified Green: The Law and Future of Environmental Labelling', *Yale Journal on Regulation* 10.1: 147-227.

Groundwork (1998) *Small Firms and the Environment: A Groundwork Status Report* (Birmingham, UK: Groundwork Foundation National Office).

Gunner, J. (1994) *Environmental Business Practices, Perceptions and Response to Regulation in Hertfordshire* (unpublished report; University of Hertfordshire, UK).

Gunningham, N. (1991) 'Private Ordering, Self-regulation and Futures Markets: A Comparative Study of Informal Social Control', *Law and Policy* 13.4: 297-326.

—— (1993) 'Environmental Auditing: Who Audits the Auditors?', *Environment and Planning Law Journal* 10.4: 229-38.

—— (1994) 'Beyond Compliance: Management of Environmental Risk', in B. Boer, R. Fowler and N. Gunningham (eds.), *Environmental Outlook: Law and Policy* (Sydney: Federation Press): 254-82.

—— (1995) 'Environment, Self-regulation, and the Chemical Industry: Assessing Responsible Care', *Law and Policy* 17.1 (January 1995): 57-109.

—— (1999) *CEO and Supervisors Drivers: Review of Literature and Current Practice* (Canberra: National Occupational Health and Safety Commission, Commonwealth of Australia).

Gunningham, N., and A. Cornwall (1994) 'Legislating the Right to Know', *Environmental Law and Planning Journal* 11.4: 274-88.

Gunningham, N., and P. Grabosky (1998) *Smart Regulation: Designing Environmental Policy* (Oxford, UK: Oxford University Press).

Gunningham, N., and R. Johnstone (1999) *Regulating Workplace Safety* (Oxford, UK: Oxford University Press): 68-94.

Gunningham, N., and J. Prest (1993) 'Environmental Audits as a Regulatory Strategy', *Sydney Law Review*, 1993: 492-526.

Gunningham, N., and J. Rees (1997) 'Industry Self-regulation: An Institutional Perspective, *Law and Policy* 19.4: 363-414.

Gunningham, N., and D. Sinclair (1997) *Barriers and Motivators to the Adoption of Cleaner Production Practices* (Canberra: Environment Australia, Commonwealth of Australia).

—— (1998) 'Next Generation Environmental Policy', *Melbourne University Law Review* 22.3: 592-616.

—— (1999a) 'Integrative Regulation: A Principle-Based Approach to Environmental Policy', *Law and Social Inquiry* 24.4: 853-96.

—— (1999b) 'Regulatory Pluralism: Designing Environmental Policy Mixes', *Law and Policy* 21: 49-76.

—— (2002) *Environmental Partnerships: Combining Commercial Advantage and Sustainability in the Agriculture Sector* (Canberra: Rural Industries Research and Development Corporation).

Gunningham, N., R. Johnstone and P. Rozen (1996) *Enforcement Measures for Occupational Health and Safety in New South Wales: Issues and Options* (Sydney: WorkCover Authority).

Gunningham, N., P. Grabosky and M. Phillipson (1999) 'Harnessing Third Parties as Surrogate Regulators: Achieving Environmental Outcomes by Alternative Means', *Business Strategy and the Environment* 8.4: 211-29.

Gunningham, N., R. Kagan and D. Thornton (forthcoming) *Shades of Green: Regulation, Business and Environment.*

Gunningham, N., D. Sinclair and P. Burritt (1998) *On the Spot Fines* (Sydney: National Occupational Health and Safety Commission, Commonwealth of Australia).

Haines, F. (1997) *Corporate Regulation* (Oxford, UK: Oxford University Press).

Hamilton, J.T. (1993) 'Pollution as News: Media and Stock Market Reactions to the TRI Data', *Journal of Environmental Economics and Management* 27.1: 38-48.

—— (1995) 'Pollution as News: Media and Stock Market Reactions to the Toxic Release Inventory Data', *Journal of Environmental Economics and Management* 28: 98-113.

Hamilton, S., S. Clay, and J. Mills (1999) 'Development of Cleaner Production Approaches Using the Automotive Repair Industry as a Case Study', presented at the *Second Asia Pacific Cleaner Production Roundtable*, United Nations Environment Programme, Brisbane (www.ens.gu.edu.au/ciep/CLEANP/CPbook/Chapt7.pdf): 371-76.

Hardin, R. (1971) 'Collective Action as an Agreeable n-Prisoners' Dilemma', *Behavioral Science* 16: 472-81.

Hardy, N. (1998) *The Altona Complex Neighbourhood Consultative Group* (Melbourne: Altona Complex Neighbourhood Consultative Group).

Harrison, K. (2001) 'Voluntarism and Environmental Governance', in E.A. Parson (ed.), *Governing the Environment: Persistent Challenges, Uncertain Innovations* (Toronto: University of Toronto Press).

Hart, S.L. (1997) 'Strategies for a Sustainable World', *Harvard Business Review* 75.1 (January/February 1997): 66-76.

Hartman, R., M. Huq and D. Wheeler (1997) *Why Paper Mills Clean Up* (Policy Research Paper WPS 1710; Washington, DC: World Bank).

Harvard Business Review on Business and the Environment (2000) *Harvard Business Review* (New York: Harvard Business School Press).

Harvard School of Public Health (2000) *Evaluation of Programa Nacional de Auditoria Ambiental (Mexico)* (unpublished; Boston, MA: Harvard School of Public Health, November 2000).

Health and Safety Executive, UK (1992) *Successful Health and Safety Management* (HS[G]65; London: HMSO).

Helms, S. (1999) 'Report Card', *Environmental Forum*, November/December 1999: 21-27.

Henriques, I., and P. Sadorsky (1995) *The Determinants of an Environmentally Responsive Firm: An Empirical Approach* (Ontario: York University, Faculty of Administrative Studies).

Herb, J., and S. Helms (2000) 'Harnessing the Power of Information: Environmental Right to Know as a Driver of Sound Environmental Policy', presented at a workshop on *Education, Information and Voluntary Measures in Environmental Protection*, National Academy of Sciences/National Research Council Committee on the Human Dimensions of Global Change, Washington, DC, November 2000.

Hietamaki, M. (2000a) 'Why to Build EIS', in M. Rapp and W. Hafner (eds.), *IMPEL 2000 Conference on Environmental Compliance and Enforcement, 11–13 Villach, Austria, October 2000: Final Report* (European Union, Environment Directorate, http://europa.eu.int/comm/environment/impel/conference_report.pdf): 71-72.

—— (2000b) 'Needs and Plans to Further Develop EIS', in M. Rapp and W. Hafner (eds.), *IMPEL 2000 Conference on Environmental Compliance and Enforcement, 11–13 Villach, Austria, October 2000: Final Report* (European Union, Environment Directorate, http://europa.eu.int/comm/environment/impel/conference_report.pdf): 76-77.

Higley, C.J., F. Convery and F. Lévêque (2001) 'Voluntary Approaches: An Introduction', presented at CAVA (Concerned Action on Voluntary Approaches) International Policy Workshop on the Use of Voluntary Approaches, Brussels, 1 February 2001, www.cerna.ensmp.fr/Progeuropeens/CAVA/Workshops.html.

Hillary, R. (ed.) (2000a) *Small and Medium-Sized Enterprises and the Environment: Business Imperatives* (Sheffield, UK: Greenleaf Publishing).

—— (2000b) 'The Eco-Management and Audit Scheme, ISO 14001 and the Smaller Firm', in R. Hillary (ed.), *Small and Medium-Sized Enterprises and the Environment: Business Imperatives* (Sheffield, UK: Greenleaf Publishing): 128-47.

Hobbs, J. (2000) 'Promoting Cleaner Production in Small and Medium-Sized Enterprises', in R. Hillary (ed.), *Small and Medium-Sized Enterprises and the Environment: Business Imperatives* (Sheffield, UK: Greenleaf Publishing): 148-57.

Holmes, C.H. (1992) *Address to Hazardous Waste Conference* (Melbourne: Australian Chemical Industry Association).

Hopkins, A. (1995) *Making Safety Work: Getting Management Commitment to Occupational Health and Safety* (Sydney: Allen & Unwin).

Howard, J., J. Nash and J. Ehrenfeld (1999) 'Industry Codes as Agents of Change: Responsible Care Adoption by US Chemical Companies', *Business Strategy and the Environment* 8: 281-95.

Howes, R., J. Skea and B. Whelan (1997) *Clean and Competitive? Environmental Performance in Industry* (London: Earthscan Publications).

Humphrey, H. (1994) 'Public/Private Environmental Auditing Agreements', *Corporate Conduct Quarterly* 3: 1.

Hutchingson, A. (1993) *Devon and Cornwall's Small and Medium Sized Enterprise Sector 'Green' Survey Results* (Plymouth, UK: Plymouth Business School).

Hutter, B. (1996) *Compliance, Regulation and Environment* (Oxford, UK: Oxford University Press).

—— (ed.) (1999) *Environmental Law* (Oxford, UK: Oxford University Press).

Industry Commission Australia (1995) *Work, Health and Safety: Inquiry into Occupational Health and Safety.* I. *Report*; II. *Appendices* (No. 47; Canberra: Commonwealth of Australia).

Information Australia (2000) *Environmental Standards Update*, 11 June 2000: 3.

ISO (International Organisation for Standardisation) (1996) *ISO 14001. Environmental Management Systems: Specification with Guidance for Use* (ISO/TC207/SC; Geneva: ISO; www.iso.ch).

Jackall, R. (1988) *Moral Mazes: The World of Corporate Managers* (New York: Oxford University Press).

Jochem, E., and W. Eichhammer (1996) 'Voluntary Agreements as an Instrument to Substitute Regulating and Economic Instruments? Lessons from the German Voluntary Agreement on CO_2 Reduction', paper presented at the conference on *Economic and Law of Voluntary Approaches to Environmental Policy*, REEM/CERNA, 18–19 November 1996, Venice.

Johannson, L. (1997) 'The Challenge of Implementing ISO 14001 for Small and Medium Sized Enterprises', *Environmental Quality Management*, Winter 1997: 9-18.

Johnstone, D. (1999) 'Foreign Direct Investment and the Environment: Challenges and Opportunities', in *OECD Proceedings Foreign Direct Investment and the Environment* (Paris: Organisation for Economic Co-operation and Development): 18.

Joskow, P.L., R. Schmalensee and E.M. Bailey (1996) *Auction Design and the Market for Sulfur Dioxide Emissions* (NBER Working Paper, W5745; Cambridge, MA: National Bureau of Economic Research, September 1996, http://papers.nber.org/papers/W5745).

Kagan, R.A. (2001) *Adversarial Legalism: The American Way of Law* (Cambridge, MA: Harvard University Press).

Kagan, R.A., and L. Axelrad (2000) *Regulatory Encounters: Multinational Corporations and Adversarial Legalism* (Berkeley, CA: University of California Press).

Karkkainen, B.C. (2000) 'Information as Environmental Regulation: TRI and Performance Benchmarking, Precursor to a New Paradigm' (Working Paper; New York: Columbia Law School).

Khanna, M., and L. Damon (1998) 'EPA's Voluntary 33/50 Program: Impacts on Toxic Releases and Economic Performance of Firms', presented at the *First World Congress of Environmental And Resource Economics*, Venice, June 1999.

Khanna, M., W. Quimio and D. Bojilova (1998) 'Toxics Release Information: A Policy Tool for Environmental Protection', *Journal of Environmental Economics and Management* 36: 243-66.

King, A., and M. Lenox (2000) 'Industry Self-regulation without Sanctions: The Chemical Industry's Responsible Care Program', *Academy of Management Journal* 43.6: 698-716.

Kleindorfer, P., and E. Orts (1996) *Informational Regulation of Informational Risks* (Working Paper; Philadelphia, PA: Wharton School, University of Pennsylvania).

Knight, A. (1994) 'International Standards for Environmental Management', *UNEP Industry and Environment* 17.3 (July 1994): 45.

Koch, C., and K. Nielsen (1996) *Working Environment Regulation: How Reflexive—How Political? A Scandinavian Case* (Working Paper; Lyngby, Denmark: Technical University of Denmark, June 1996).

Konar, S., and M. Cohen (1997) 'Information as Regulation: The Effect of Community Right to Know Laws on Toxic Emissions', *Journal of Environmental Economics and Management* 32: 109-24.

Koski, O. (2000) 'Presentation on the Finnish VAHTI-System: Database, Emission, Discharge, Waste, Air and Water Quality, Monitoring', in M. Rapp and W. Hafner (eds.), *IMPEL 2000 Conference on Environmental Compliance and Enforcement, 11–13 Villach, Austria, October 2000: Final Report* (European Union, Environment Directorate, http://europa.eu.int/comm/environment/impel/conference_report.pdf): 74.

KPMG (1996) *Canadian Environmental Management Survey* (Toronto: KPMG).

—— (1997) *The Environmental Challenge and Small and Medium Sized Enterprises in Europe* (The Hague: KPMG Consulting).

Krarup, S. (2001) 'Can Voluntary Approaches be Environmentally Effective and Economically Efficient?', presented at CAVA (Concerned Action on Voluntary Approaches) International Policy Workshop on the Use of Voluntary Approaches, Brussels, 1 February 2001, www.cerna.ensmp.fr/Progeuropeens/ CAVA/Workshops.html.

Lanoie, P., B. Laplante and M. Roy (1997) *Can Capital Markets Create Incentives for Pollution Control?* (Washington, DC: World Bank Policy Research Department).

Leyden, P. (1997) 'Trading in Southern California', presented to American Bar Association 26th Annual Conference on Environmental Law, Keystone, CO, March 1997.

Long, F.J., and M.B. Arnold (1994) *The Power of Environmental Partnerships* (Fort Worth, TX: Dryden Press).

Lonti, Z., and A. Verma (1999) 'Industry Self-Management as a Strategy for Restructuring Government: The Case of the Ministry of Consumer and Commercial Relations (MCCR) and the Technical Standards and Safety Authority (TSSA) in Ontario' (Discussion Paper, W07; Human Resources in Government Series; Ottawa: Canadian Policy Research Networks, December 1999, www.cprn.org/docs/ work/cfg_e.pdf).

Lyon, T., and J. Maxwell (1999) 'Voluntary Approaches to Environmental Regulation: A Survey', in M. Franzine and A. Nicita (eds.), *Environmental Economics, Past, Present and Future* (Aldershot, UK: Ashgate Publishing).

Marcus, A.A. (1999) *Unlocking a New Competence for Environmental Management: Learning from Minnesota's Project Excellence in Leadership* (Report to the Environmental Protection Agency, USA).

Mazurek, J. (2000) 'Government-Sponsored Voluntary Programs in Firms: An Initial Survey', presented at National Academy of Social Sciences/National Research Council workshop on Education, Information and Voluntary Measures in Environment Protection, November 2000.

MCA (Minerals Council of Australia) (2000) *Australian Minerals Industry Code for Environmental Management: 'Backgrounder'* (Melbourne: MCA, June 2000).

McCloskey, J., and S. Maddock (1994) 'Environmental Management: Its Role in Corporate Strategy', *Management Decision* 32.1: 27-32.

Menell, P. (1996) 'Educating Consumers about the Environment: Labels versus Prices', in E. Eide and R. van den Bergh (eds.), *Law and Economics of the Environment* (Oslo: Juridisk Forlag).

Merritt, J. (1998) 'EMS in SME Don't Go? Attitudes, Awareness and Practices in the London Borough of Croydon', *Business Strategy and the Environment* 7: 90-100.

Metzenbaum, S. (2000) 'Information Driven', *Environmental Forum,* March/April 2000: 26-36.

Miller, A.S. (1994) 'The Origins and Current Directions of United States Environmental Law and Policy: An Overview', in B. Boer, R. Fowler and N. Gunningham (eds.), *Environmental Outlook: Law and Policy* (Sydney: Federation Press).

Mining Journal (1999) 'Focus and Comment: "Earning a Social Licence" ', *The Mining Journal,* 11 June 1999: 441.

Mitchell, R.B. (1994) *Intentional Oil Pollution at Sea: Environmental Policy and Treaty Compliance* (Cambridge, MA: The MIT Press).

Moffet, J., and F. Bregha (1999) 'An Overview of Issues with Respect to Voluntary Environmental Agreements', *Journal of Environmental Law and Practice* 8.1 (December 1999): 63-94.

Mol, A.P.J. (1995) *The Refinement of Production* (Utrecht, Netherlands: Van Arkel).

Mol, A.P.J., and D.A. Sonnenfeld (2000) 'Ecological Modernisation around the World: An Introduction', *Environmental Politics* 1 (Spring 2000): 3-14.

Moynihan, R. (2000) 'Organics a Go-Go', *Australian Financial Review,* 1 November 2000: 2.

Mullin, R. (1992) 'Canadian Deadline Approaches: Contemplating Continuous Improvement', *Chemical Week,* 17 June 1992: 28.

Murphy, D., and J. Bendell (1997) *In the Company of Partners: Business, Environmental Groups and Sustainable Development Post-Rio* (Bristol, UK: Policy Press).

NAPA (US National Academy of Public Administration) (1997) *Resolving the Paradox of Environmental Protection: An Agenda for Congress, EPA and the States* (Washington, DC: NAPA).

Nash, J. (2000) 'Voluntary Codes of Practice: Non-Governmental Institutions for Promoting Environmental Management in Firms', presented at a workshop on *Education, Information and Voluntary Measures in Environmental Protection,* National Academy of Sciences/National Research Council Committee on the Human Dimensions of Global Change, Washington, DC, November 2000.

Newton, T., and G. Harte (1997) 'Green Business: Technicist Kitsch?', *Journal of Management Studies* 34.1 (January 1997): 75-98.

NRC (US National Research Council) (1997) *Fostering Industry-Initiated Environmental Protection Efforts* (Washington, DC: National Academy Press).

O'Connor, J. (1996) 'The Second Contradiction of Capitalism', in T. Benton (ed.), *The Greening of Marxism* (New York: Guilford).

OECD (Organisation for Economic Co-operation and Development) (1991) *Environmental Labelling in OECD Countries* (Paris: OECD).

—— (1995) *Technologies for Cleaner Production and Products: Towards Technological Transformation for Sustainable Development* (Paris: OECD).

—— (1997) 'Greener Public Purchasing', Issues Paper Prepared for 'Green Goods', 4th International Conference on Greener Public Purchasing, Biel-Bierne, Switzerland, 24–26 February 1997.

—— (1998) *Putting Markets to Work: The Design and Use of Marketable Permits and Obligations* (Paris: OECD).

—— (1999) *Voluntary Approaches for Environmental Policy: An Assessment* (Paris: OECD).

—— (2000a) *Reducing the Risk of Policy Failure: Challenges for Regulatory Compliance* (Working Paper 77; Paris: OECD/PUMA).

—— (2000b) *Private Initiatives for Corporate Responsibility: An Analysis* (Paris: OECD DAFFE [Directorate for Financial, Fiscal and Enterprise Affairs]/IME [Committee on International Investment and Multinational Enterprises]).

—— (2000c) *Corporate Responsibility: Results of a Fact Finding Mission on Private Initiatives* (Paris: OECD DAFFE/IME [2000] 15).

—— (2000d) *Public Policy and Voluntary Initiatives: What Roles have Governments Played?* (Paris: OECD DAFFE/IME 15/ANN5).

—— (2001) *Voluntary Initiatives for Corporate Responsibility: Progress to Date* (provisional title) (Paris: OECD).

Olivecrona, C. (1995a) 'The Nitrogen Oxide Charge on Energy Production in Sweden', in R. Gale, S. Barg and A. Gilles (eds.), *Green Budget Reform* (London: Earthscan Publications): 163-72.

—— (1995b) 'Carbon Dioxide Taxes in Scandinavia', in R. Gale, S. Barg and A. Gilles (eds.), *Green Budget Reform* (London: Earthscan Publications): 173-84.

O'Rourke, D. (2000) *Monitoring the Monitors: A Critique of PricewaterhouseCoopers (PwC) Labor Monitoring* (Cambridge, MA: MIT University, Department of Urban Studies and Planning).

Orts, W.E. (1995) 'Reflexive Environmental Law', *Northwestern University Law Review* 89.4: 1227-1340.

Osborne, D., and T. Gaebler (1992) *Reinventing Government: How the Entrepreneurial Spirit is Transforming the Public Sector* (Boston, MA: Addison-Wesley).

Owen, D. (1996) 'A Critical Perspective on the Development of European Corporate Environmental Accounting and Reporting', presented at the 1996 Australian Academy of Science *Fenner Conference on the Environment: Linking Environment and Economy through Indicators and Accounting Systems*, University of New South Wales, Sydney, 30 September–3 October 1996.

Pacific and Yukon Regional Office of Environment Canada (1998) *Enforcement vs Voluntary Compliance: An Examination of the Strategic Enforcement Initiatives Implemented by the Pacific and Yukon Regional Office of Environment Canada 1983 to 1998* (Report no. DOE FRAP 19983; North Vancouver, BC: Environment Canada).

Panayotou, T. (1998) *Instruments of Change: Motivating and Financing Sustainable Development* (London: Earthscan Publications).

Parker, C. (1999a) 'Compliance Professionalism and Regulatory Community: The Australian Trade Practices Regime', *Journal of Law and Society* 26: 215-39.

—— (1999b) *Just Lawyers* (Oxford, UK: Oxford University Press).

—— (1999c) 'Evaluating Regulatory Compliance: Best Practice and Standards', *Trade Practices Law Journal* 7.2: 62-71.

—— (2000) 'Summary of the Scholarly Literature on Regulatory Compliance', Annex to *Reducing the Risk of Policy Failure: Challenges for Regulatory Compliance* (Working Paper 77; Paris: Organisation for Economic Co-operation and Development).

Pearson, C.S. (2000) *Economics in the Global Environment* (Cambridge, UK: Cambridge University Press).

Petts, J. (2000) 'Small and Medium-Sized Enterprises and Environmental Compliance: Attitudes among Management and Non-management', in R. Hillary (ed.), *Small and Medium-Sized Enterprises and the Environment: Business Imperatives* (Sheffield, UK: Greenleaf Publishing): 49-60.

Poncelet, E. (1999) 'In Search of the "Win–Win": Possibilities and Limitations of Multi-stakeholder Environmental Partnerships', *Greening of Industry Conference, Best Paper Proceedings*.

—— (forthcoming) 'A Kiss Here and a Kiss There: Conflict and Collaboration in Environmental Partnerships', *Environmental Management* 27: 13-25.

Porter, M. (1991) 'America's Green Strategy', *Scientific American* 264 (April 1991): 8.

—— (1998) 'How Competitive Forces Shape Strategy', in H. Mintzberg, J.B. Quinn and S. Goshal (eds.), *The Strategy Process* (Bath, UK: Prentice Hall): 60-69.

Porter, M., and C. van der Linde (1995) 'Green and Competitive: Ending the Stalemate', *Harvard Business Review*, September/October 1995: 120-34.

Posner, T. (1992) *The Engineer*, 5 March 1992: 20.

Priest, M. (1998-99) 'The Privatization of Regulation: Five Models of Self-Regulation', *Ottawa Law Review* 29: 233-302.

Purdue, M. (1991) 'Integrated Pollution Control in the Environmental Protection Act 1990: A Coming of Age of Environmental Law?', *Modern Law Review* 54: 538-39.

Quality Environmental Management Sub-Committee, President's Commission on Environmental Quality (1993) *Total Quality Management: A Framework for Pollution Reduction* (Washington, DC: President's Commission on Environmental Quality, January 1993).

Queensland Department of Employment, Vocational Educational, Training and Industrial Relations (1995) *Work: Health and Safety*, Vol. II (Report No. 47; Canberra: AGPS)

Ramesohl, S., and K. Kristof (1999) *A Socio-economic Analysis of Energy-Related Voluntary Agreements in Germany: Transactions Costs and Innovations* (CAVA Working Paper; Paris: CERNA).

Ransom, P., and D. Lober (1999) 'Why do Firms set Environmental Performance Goals? Some Evidence from Organisational Theory', *Business Strategy and the Environment* 8: 1-13.

Rapp, M., and W. Hafner (eds.) (2000) *IMPEL 2000 Conference on Environmental Compliance and Enforcement, 11–13 Villach, Austria, October 2000: Final Report* (European Union, Environment Directorate, http://europa.eu.int/comm/environment/impel/conference_report.pdf).

Rees, J.V. (1994) *Hostages of Each Other: The Transformation of Nuclear Safety Since Three Mile Island* (Chicago: University of Chicago Press).

—— (1997) 'The Development of Communitarian Regulation in the Chemical Industry', *Law and Policy* 19.4: 477-528.

—— (1988) *Reforming the Workplace: A Study of Self-regulation in Occupational Safety* (Philadelphia, PA: University of Pennsylvania Press).

Rehbinder, E. (1993) 'Environmental Regulation through Fiscal and Economic Incentives in a Federalist System', *Ecology Law Quarterly* 20: 57-83.

Reilly, W. (1990) 'Aiming before we Shoot: The Quiet Revolution in Environmental Policy', address to the National Press Club, Washington, DC, 26 September 1990.

Reinhardt, R. (2000) *Down to Earth* (Cambridge, MA: Harvard Business School Press).

Rimington, R. (1998) *Managing Risk—Adding Value* (London: Health and Safety Executive, HMSO).

Rogers, Jr, J.E. (1992) 'Adopting and Implementing a Corporate Environmental Charter', *Business Horizons* 35.2: 29-33.

Rose, C.M. (1997) 'Property Rights and Responsibilities', in M. Chertow and D. Esty (eds.), *Thinking Ecologically: The Next Generation of Environmental Policy* (New York: Yale University Press): 49-59.

Sabel, C., A. Fung and B. Karkkainen (2000) *After Backyard Environmentalism* (Boston, MA: Beacon Press).

Schmidheiny, S (1992) *Changing Course: A Global Business Perspective on Development and the Environment* (Cambridge, MA: The MIT Press).

Schmidheiny, S., and F. Zorraquín (1996) *Financing Change: The Financial Community, Eco-Efficiency and Sustainable Development* (Cambridge, MA: The MIT Press).

Schnutenhaus, J.O. (1995) 'Tax Differentials for Catalytic Converters and Unleaded Petrol in Germany', in R. Gale, S. Barg and A. Gilles (eds.), *Green Budget Reform* (London: Earthscan Publications): 79-90.

Scholz, J.T. (1984) 'Cooperation, Deterrence, and the Ecology of Regulatory Enforcement', *Law and Society Review* 18: 179-224.

Segerson, K., and T. Miceli (1998) 'Voluntary Environmental Agreements: Good or Bad News for Environmental Protection?', *Journal of Environmental Economics and Management* 38: 158-75.

Seidenfeld, M. (2000) 'An Apology for Administrative Law in "The Contracting State"', *Florida State University Law Review* 28 (Fall 2000).

Simon, H. (1992) *Economics, Bounded Rationality and the Cognitive Revolution* (Cheltenham, UK: Edward Elgar).

Sinclair, Knight, Merz (2001) *AMEEF Research Project Industry Based Initiatives: A Brief Progress Report* (Sydney: Sinclair, Knight, Merz, 6 July 2001).

Smart, B. (ed.) (1992) *Beyond Compliance: A New Industry View of the Environment* (Washington, DC: World Resources Institute).

Smithers, R. (1989) 'Chemical firms adopt code to clean up the industry', *The Age*, 27 September 1989: 5.

Solomon, F. (2000) 'External Verification of the Australian Minerals Industry Code for Environmental Management: A Case Study', *Australian Journal of Environmental Management* 7 (June 2000): 91-98.

Sparrow, M. (2000) *The Regulatory Craft* (Washington, DC: Brookings Institution).

Spillman, P. (1998) 'The Western Australian Environment Protection Amendment Act 1998', *Journal of Environmental Law and Policy*, August 1998: 287-92.

Stavins, R., and B. Whitehead (1997) 'Market Based Environmental Policies', in M. Chetow and D. Esty (eds.), *Thinking Ecologically: The Next Generation of Environmental Policy* (New Haven, CT: Yale University Press).

Steinzor, R. (1996) 'Regulating Reinvention: Does the emperor have any clothes?', *Environmental Law Reporter, News and Analysis*, 1996.

—— (1998) 'Reinventing Environmental Regulation: The Dangerous Journey from Command to Self-Control', *Harvard Environmental Law Review* 22: 103

Stratos and Pollution Probe (2000) *Reinforcing the Business Case for Environmental Voluntary Initiatives* (Ottawa: Environment Canada, May 2000).

Stromvag, A. (1998) 'How an OHS Management System for Contractors can Improve Profitability', *Journal of Occupational Health and Safety, Australia/New Zealand* 14.5: 505-509.

Sunstein, C. (1990) 'Paradoxes of the Regulatory State', *University of Chicago Law Review* 57: 408-27.

Susskind, L., and J. Seconda (1998) 'The Risks and Advantages of Agency Discretion: Evidence from the EPA's Project XL', *UCLA Journal of Environmental Law and Policy* 17: 67-116.

Swift, B. (1997) 'The Acid Rain Test', *Environmental Forum*, May/June 1997: 17-25.

Teubner, G. (1983) 'Substantive and Reflexive Elements in Modern Law', *Law and Society Review* 17: 239.

Teubner, G., L. Farmer and D. Murphy (eds.) (1994) *Environmental Law and Ecological Responsibility: The Concept and Practice of Ecological Self-organization* (Chichester, UK: John Wiley).

Tibor, T., and I. Feldman (1996) *ISO 14000: A Guide to the New Environmental Management Standards* (Chicago: Irwin).

Tilley, F. (1999) 'The Gap between Environmental Attitude and Environmental Behaviour of Small Firms', *Business Strategy and the Environment* 8: 241.

—— (2000) 'Small Firms' Environmental Ethics: How deep do they go?', in R. Hillary (ed.), *Small and Medium-Sized Enterprises and the Environment: Business Imperatives* (Sheffield, UK: Greenleaf Publishing): 35-48.

Tsutsumi, R. (1999) 'The Nature of the Voluntary Agreement in Japan' (CAVA Working Paper, 99/10/8, www.akf.dk/cava/paper/abstract/tsutsumi.htm).

UK Round Table on Sustainable Development (1999) *Small and Medium-Sized Enterprises*, www.open.gov.uk/roundtbl/hometb.htm

UNEP (United Nations Environment Programme) (1992) *Voluntary Industry Codes of Conduct for the Environment* (Technical Report 40; Geneva: UNEP).

Unglik, A. (1996) *Between a Rock and a Hard Place* (Melbourne: Environment Protection Agency).

US EPA (Environmental Protection Agency) (1994) *Sustainable Industry: Promoting Strategic Environmental Protection in the Industrial Sector* (Phase 1 Report; Washington, DC: EPA).

—— (1996) *EPA Acid Rain Program* (Update No. 3, Technology and Information; Washington, DC: EPA).

—— (1999) *Aiming for Excellence* (Washington, DC: EPA, Office of the Administrator).

—— (2000) *A Decade of Progress: Innovation at the Environmental Protection Agency* (Washington, DC: EPA Innovation Annual Report).

US General Accounting Office (1991) *Toxic Chemicals: EPA's Toxic Release Inventory is Useful but can be Improved* (GAO/RCED-91-121; Washington, DC: GAO).

—— (1997) *Global Warming: Information on the Results of Four of EPA's Voluntary Climate Change Programs* (GAO/RCED-97-163; Washington, DC: GAO, 30 June 1997).

VACC (Victorian Automotive Chamber of Commerce)/VEPA (Victorian Environment Protection Agency) (undated) *Environmental Management: It's an Investment. A Clean Green Shop Brochure* (Melbourne: VACC/VEPA).

VEPA (Victorian Environment Protection Agency) (1993a) *A Question Of Trust: Accredited Licensee Concept* (Discussion Paper, Publication 285; Melbourne: VEPA, July 1993)

—— (1993b) *EPA Information Bulletin: Environment Improvement Plans* (Publication 394; Melbourne: VEPA, October 1993).

—— (1994) *EPA Information Bulletin: Accredited Licensee Guidelines for Applicants* (Publication 424; Melbourne: VEPA, October 1994).

—— (1995) *EPA Information Bulletin* (Melbourne: VEPA, July 1995).

—— (1997) *Cleaner Production Case Studies* (Publication 536; Melbourne: VEPA).

—— (1998) *Information Bulletin: Accredited Licence System, Guidelines for Environmental Management System Certification* (Melbourne: VEPA, June 1998, www.epa.vic.gov.au).

—— (1998–2001) *EPA Corporate Plan* (Melbourne: VEPA).

—— (undated) *Technical Guideline 201* (Melbourne: VEPA).

Victorian Legislative Council (1989) *Environment Protection (General Amendment) Bill* (second reading speech, 17 November 1989).

Waters and Rivers Commission (1999) *Swan–Canning Industry Survey, Pilot Study Findings* (Perth, Australia: Swan River Trust, Waters and Rivers Commission, December 1999).

WBCSD (World Business Council for Sustainable Development) (1995) *Eco-Efficient Leadership* (Geneva: WBCSD).

Webb, K. (1996) 'Voluntary Codes: A Synopsis', in Office of Consumer Affairs, *Voluntary Codes Symposium* (Ottawa: Office of Consumer Affairs, Industry Canada and Regulatory Affairs, Treasury Board Canada).

—— (ed.) (2002) *Voluntary Codes: Private Governance, the Public Interest and Innovation* (Ottawa: Carleton University Research Unit for Innovation, Science and Environment).

Webb, K., and A. Morrison (1996) 'The Legal Aspects of Voluntary Codes', in Office of Consumer Affairs, *Voluntary Codes Symposium* (Ottawa: Office of Consumer Affairs, Industry Canada and Regulatory Affairs, Treasury Board Canada).

Wells, R.C. (2001) 'Voluntary Initiatives for the Minerals Industry: The Australian Experience', presented at the Mining, Minerals and Sustainable Development workshop, Santa Fe, CA, July 2001.

Wells, R., and D. Galbraith (1998) *The Guadalajara Environmental Management Pilot Improving the Environmental Management Capabilities of Small and Medium Companies in Guadalajara, Mexico* (Lexington, MA: Lexington Group): 35.

Wills, I. (2000) *Industry–Community–EPA Consultation in Pollution Control: The Victorian Experience* (Melbourne: Monash University, Department of Economics).

—— (2001) 'Information Exchange, Risk and Community Participation in Pollution Control Measures', paper presented at a seminar sponsored by the Hillman Endowment, University of Arizona, Tucson, AZ, 19 May 2000.

Wills, I., and S. Fritschy (2001) 'Industry–Community Regulator Consultation in Improving Environmental Performance in Victoria', *Australian Journal of Environmental Management* 8.3: 158.

Wisconsin Department of Natural Resources (2000) *Wisconsin's Green Tier Regulatory Proposal* (www.dnr.state.wi.us/org/caer/cea/green_tier/factsheets/proposal.htm, January 2000).

Work Life (1999) *Occupational Health and Safety Management Systems in Small Companies Workshop Summary* (Stockholm: Work Life, June 1999).

Wright, M. (1998) *Factors Motivating Proactive Health and Safety Management* (Contract Research Report prepared by Entec UK Ltd for the Health and Safety Executive; London: HMSO).

WWF (World Wide Fund for Nature) (2000) *Ore or Overburden II? WWF's 2nd Annual Scorecard on Mining Company Environmental Reports* (Melbourne: WWF Australia, September 2000).

WWF (World Wide Fund for Nature)–Placer Dome Asia Pacific (2001) *Mining Certification Evaluation Project* (Discussion Paper; WWF Australia Resource Conservation Program, Mineral Resources Unit, January 2001).

Zarsky, L. (1999) 'Havens, Halos and Spaghetti: Untangling the Evidence about Foreign Direct Investment and the Environment', in *OECD Proceedings, Foreign Direct Investment and the Environment* (Paris: Organisation for Economic Co-operation and Development).

Zeimet, D., D. Ballard and K. Mai (1997) 'A Comprehensive Safety and Health Program for the Small Employer', *Occupational Health and Safety* 66.10 (October 1997): 127-33.

ABBREVIATIONS

ABS	Australian Bureau of Statistics
ACIC	Australian Chemical Industry Council
ACNCG	Altona Complex Neighbourhood Consultative Group
AIIIC	Automotive Industry's International Information Center
AMICEM	Australian Minerals Industry Code for Environmental Management
APEC	Asia Pacific Economic Co-operation
ARET	Accelerated Reduction and Elimination of Toxics (Canada)
CEC	Commission of the European Communities
CEO	chief executive officer
CERES	Coalition of Environmentally Responsible Economies
CFC	chlorofluorocarbon
CLC	Community Liaison Committee
CPPP	Cleaner Production Partnerships Program
CRTK	community right-to-know
DEP	Department of Environmental Protection (Western Australia)
EAP	environment audit programme
EDF	Environmental Defense Fund
EEA	European Environment Agency
EIP	Environmental Improvement Plan
EMAS	Eco-management and Audit Scheme
EMP	European Member of Parliament
EMS	environmental management system
ENDS	Environmental Data Services
EPA	Environmental Protection Agency
EPER	European Pollutant Emission Register
EU	European Union
EUROP	European Retailer Produce Working Group
FDI	foreign direct investment
FSC	Forest Stewardship Council
GDP	gross domestic product
GEMI	Global Environmental Management Initiative
GRI	Global Reporting Initiative
ICME	International Council on Metals and the Environment
INPO	Institute of Nuclear Power Operations (USA)
IPPC	Integrated Pollution Prevention and Control
ISO	International Organisation for Standardisation
LEAF	Linking Environment and Farming

MARPOL	International Convention for Prevention of Pollution by Ships
MCA	Minerals Council of Australia
MNE	multinational enterprise
MOE	Ministry of Environment (British Columbia)
MOU	Memorandum of Understanding
MPCA	Minnesota Pollution Control Agency
MSWG	Multi-State Working Group on the Environment (USA)
NAPA	National Academy of Public Administration (USA)
NCAP	National Community Advisory Panel
NGO	non-governmental organisation
NIMBY	not in my back yard
NPI	National Pollutant Inventory (Australia)
NRC	Nuclear Regulatory Commission (USA)
NRC	National Research Council (USA)
NSW	New South Wales
OECD	Organisation for Economic Co-operation and Development
OHS	occupational health and safety
PACIA	Plastics and Chemicals Industries Association Inc. (Australia)
PIM	Printing Industry of Minnesota Inc.
PROPER	Programme for Pollution Control, Evaluation and Rating (Indonesia)
PROKASIH	Program Kali Bersih (Indonesia)
PUMA	Public Management and Governance (OECD)
QA	quality assurance
RACIRB	Refrigeration and Air Conditioning Registration Board
SBT	segregated ballast tank
SCEEMAS	Small Company Environmental and Energy Management Assistance Scheme (UK)
SCRAMS	safety-related automatic shutdown
SME	small or medium-sized enterprise
SO_2	sulphur dioxide
TAFE	Technical and Further Education
TMI	Three Mile Island
TRI	Toxics Release Inventory (USA)
UNEP	United Nations Environment Programme
UST	underground storage tank
VA	voluntary agreement
VACC	Victorian Automotive Chamber of Commerce
VAHTI	Finnish compliance monitoring system
VEPA	Victorian Environment Protection Authority
VVGA	Victorian Vegetables Growers' Association
WBCSD	World Business Council for Sustainable Development
WWF	World Wide Fund for Nature

INDEX